AF411349

Applications of Internet of Things

Applications of Internet of Things

Editors

Chi-Hua Chen
Kuen-Rong Lo

MDPI • Basel • Beijing • Wuhan • Barcelona • Belgrade • Manchester • Tokyo • Cluj • Tianjin

Editors
Chi-Hua Chen
College of Mathematics and
Computer Science,
Fuzhou University
China

Kuen-Rong Lo
Telecommunication Laboratories,
Chunghwa Telecom Co. Ltd.
Taiwan

Editorial Office
MDPI
St. Alban-Anlage 66
4052 Basel, Switzerland

This is a reprint of articles from the Special Issue published online in the open access journal *ISPRS International Journal of Geo-Information* (ISSN 2220-9964) (available at: https://www.mdpi.com/journal/ijgi/special_issues/Internet_things_2016).

For citation purposes, cite each article independently as indicated on the article page online and as indicated below:

LastName, A.A.; LastName, B.B.; LastName, C.C. Article Title. *Journal Name* **Year**, *Volume Number*, Page Range.

ISBN 978-3-0365-1192-4 (Hbk)
ISBN 978-3-0365-1193-1 (PDF)

Contents

About the Editors

Chi-Hua Chen is a distinguished professor of Fuzhou University and a chair professor of Dalian Maritime University. He received his Ph.D. degree from National Chiao Tung University (NCTU) in 2013. Furthermore, he served as an assistant professor of National Tsing Hua University, NCTU, National Taipei University, and National Kaohsiung University of Science and Technology. He also served as a research fellow for the Telecommunication Laboratories of Chunghwa Telecom Co. Ltd. He has published over 300 academic articles and 50 patents. Some of these academic articles were published in IEEE Internet of Things Journal, IEICE Transactions, etc. He has hosted several projects which were funded by the National Natural Science Foundation of China, Fujian Province, etc. He serves as an editor for several SCI-indexed journals (e.g., Scientific Data (one of the Nature Research Journals), IEICE Transactions on Information and Systems, IEEE Access, etc.). He also serves as a chair for several international conferences (e.g., WWW 2021 Workshop, DASFAA 2021 Workshop, IEEE BIBM 2020 Workshop, IEEE TrustCom 2020 Workshop, etc.). His recent research interests include Internet of Things, machine learning and deep learning, mobile communications, and intelligent transportation systems.

Kuen-Rong Lo is an IoT (Internet of Things) Laboratory Managing Director of the Telecommunication Laboratories, Chunghwa Telecom Co. Ltd. He received his Ph.D. degree from National Chiao Tung University (NCTU) in 2000. From 1984 to 2018, he was with Telecommunication Laboratories, as a Senior Researcher, Division Project Leader, Project Manager, and then Managing Director. He was involved in designing IoT key technologies such as device management, connection management and wireless communication. From 2019 to 2021, he was a Managing Director of Data Communications Business Group and Telecommunication Laboratories, Chunghwa Telecom Co. Ltd. There, he won Chunghwa Telecom's Award for Outstanding Personnel and Outstanding Innovation Team, respectively. His recent research interests include AI-based energy management solutions, clean energy, intelligent transportation systems, digital health and big data and analytics.

 International Journal of
Geo-Information

Editorial

Applications of Internet of Things

Chi-Hua Chen [1,*] and Kuen-Rong Lo [2]

[1] College of Mathematics and Computer Science, Fuzhou University, Fuzhou 350116, China
[2] Telecommunication Laboratories, Chunghwa Telecom Co. Ltd., Taoyuan 326, Taiwan; lo@cht.com.tw
* Correspondence: chihua0826@gmail.com; Tel.: +886-975292259

Received: 17 August 2018; Accepted: 20 August 2018; Published: 22 August 2018

Abstract: This editorial introduces the special issue entitled "Applications of Internet of Things", of ISPRS International Journal of Geo-Information. Topics covered in this issue include three main parts: (I) intelligent transportation systems (ITS), (II) location-based services (LBS), and (III) sensing techniques and applications. Three papers on ITS are as follows: (1) "Vehicle positioning and speed estimation based on cellular network signals for urban roads," by Lai and Kuo; (2) "A method for traffic congestion clustering judgment based on grey relational analysis," by Zhang et al.; and (3) "Smartphone-based pedestrian's avoidance behavior recognition towards opportunistic road anomaly detection," by Ishikawa and Fujinami. Three papers on LBS are as follows: (1) "A high-efficiency method of mobile positioning based on commercial vehicle operation data," by Chen et al.; (2) "Efficient location privacy-preserving k-anonymity method based on the credible chain," by Wang et al.; and (3) "Proximity-based asynchronous messaging platform for location-based Internet of things service," by gon Jo et al. Two papers on sensing techniques and applications are as follows: (1) "Detection of electronic anklet wearers' groupings throughout telematics monitoring," by Machado et al.; and (2) "Camera coverage estimation based on multistage grid subdivision," by Wang et al.

Keywords: internet of things; intelligent transportation systems; location-based services; sensing techniques and applications

---

1. Introduction

In recent years, the techniques of Internet of Things (IoT) and mobile communication have been developed to detect human and environment information (e.g., geo-information [1,2], weather information [3,4], bio-information [5,6], human behaviors [7,8], etc.) for a variety of intelligent services and applications. The three layers in IoT are sensor, networking, and application layers [9–11]. For sensor and networking layers, the rise of mobile technology advancements [12–15] (e.g., wireless sensor networking, Wi-Fi, Bluetooth, smart mobile device, and Long Term Evolution (LTE)) has led to a new wave of machine-to-machine (M2M), machine-to-human (M2H), human-to-human (H2H), and human-to-machine (H2M) communications [16–20]. For the application layer, several IoT applications, which include energy [21,22], enterprise [23,24], healthcare [25,26], public services [27,28], residency [29,30], retail [31,32], and transportation [33,34], have been designed and implemented to detect environmental changes and send instant updates to a cloud computing server farm via mobile communications and middleware for big geo-data analyzes [35,36]. For instance, on-board units in cars can instantly detect and share information about the geolocation of the car, speed, following distance, and gaps with other neighboring cars [37–40]. While the area of IoT applications and mobile communication is a rapidly expanding field of scientific research, several open research questions still need to be discussed and studied. Therefore, the aim of this special issue is to introduce the readers a number of papers on various aspects of IoT applications.

This special issue has received a total of 23 submitted papers with only 8 papers [41–48] accepted. A high rejection rate of 65.21% of this issue from the review process is to ensure that high-quality papers with significant results are selected and published. The statistics of the special issue is presented as follows:

- Submissions (23);
- Publications (8);
- Rejections (15).

The distribution of authors' country is showed as follows:

- China (5);
- Korea (2);
- Brazil (1);
- Chile (1);
- Japan (1);
- Spain (1).

Topics covered in this issue include three main parts: (1) intelligent transportation systems (ITS), (2) location-based services (LBS), and (3) sensing techniques and applications. The three topics and accepted papers are briefly described below.

2. Intelligent Transportation Systems

Three papers on ITS are as follows: (1) "Vehicle positioning and speed estimation based on cellular network signals for urban roads," by Lai and Kuo [41]; (2) "A method for traffic congestion clustering judgment based on grey relational analysis," by Zhang et al. [42]; and (3) "Smartphone-based pedestrian's avoidance behavior recognition towards opportunistic road anomaly detection," by Ishikawa and Fujinami [43].

Lai and Kuo from China in "Vehicle positioning and speed estimation based on cellular network signals for urban roads" proposed a vehicle positioning method and a speed estimation method to analyze the cell IDs, cell sequences, and the cell dwell time of connected cells from cellular floating vehicle data (CFVD). The cell sequences can be considered to support the analysis of the judgment of urban road direction, and the cell dwell time of connected cells can be considered to support the analysis of the discrimination of proximal urban roads. The location and vehicle speed can be estimated by the k-nearest neighbor algorithm in accordance with the CFVD. In experimental environments, six urban road segments in Kaohsiung and Pingtung in Taiwan were driven in 27 runs for the evaluation of the proposed methods, and the results showed that the accuracies of vehicle positioning and speed estimation were 100% and 83.81%, respectively [41].

Zhang et al. from China and Chile in "A method for traffic congestion clustering judgment based on grey relational analysis" proposed a grey relational membership degree rank clustering algorithm based on a grey relational clustering model to analyze the traffic information (e.g., traffic flow velocity, traffic flow density and traffic volume) for the detection of traffic congestion. The proposed method based on grey relational analysis can obtain the membership degree rank of classes for judging the rank of data objects and improving the accuracy of traffic congestion detection. In experimental environments, the practical traffic flow records were collected from 30 drivers to evaluate the proposed method, and the results showed that the average accuracy of the proposed algorithm was 24.9% higher than that of the K-means algorithm [42].

Ishikawa and Fujinami from Japan in "Smartphone-based pedestrian's avoidance behavior recognition towards opportunistic road anomaly detection" used a random forest method as the classifier to analyze the azimuth patterns from smartphones for the detection of pedestrians' avoidance behaviors, and the road anomalies can be detected in accordance with the pedestrians' avoidance behaviors. In experimental environments, the practical pedestrians' avoidance behaviors were collected from 7 males and 2 females to evaluate the proposed method, and the results showed that the average accuracy of the proposed method was higher than that of other methods [43].

3. Location-Based Services

Three papers on LBS are as follows: (1) "A high-efficiency method of mobile positioning based on commercial vehicle operation data," by Chen et al. [44]; (2) "Efficient location privacy-preserving k-anonymity method based on the credible chain," by Wang et al. [45]; and (3) "Proximity-based asynchronous messaging platform for location-based Internet of things service," by gon Jo et al. [46].

Chen et al. from China in "A high-efficiency method of mobile positioning based on commercial vehicle operation data" proposed a mobile positioning method to analyze the information of global positioning system (GPS) and cellular network signals from commercial vehicle operation data for estimating the location of each cell-RSSI (received signal strength indication) pair in training stage. In the runtime stage, the trained location of each cell-RSSI pair was used to estimate the location of the vehicle in accordance with the information of cell and RSSI for mobile positioning. In experimental environments, 6,571,550 practical commercial vehicle operation records were collected to evaluate the proposed method, and the results showed that the average location error of the proposed method was lower than cell ID-based method [44].

Wang et al. from China in "Efficient location privacy-preserving k-anonymity method based on the credible chain" analyzed the user's environment and social attributes to determine the optimal k value for a k-anonymous location privacy protection method, and the k location nodes were contained in a fake trajectory which can be generated based on the credible chain. In experimental environments, numerical analysis and simulations were given to evaluate the proposed method, and the results showed that the service accuracy of the proposed method was 100% [45].

Gon Jo et al. from Korea in "Proximity-based asynchronous messaging platform for location-based Internet of things service" proposed a distance-based asynchronous messaging platform based on a location-based message-delivery protocol. The proposed platform and protocol can be used to disperse traffic and improve stability. In experimental environments, the proposed platform and protocol were implemented to analyze the transmission time and response time for the verification of obtaining location-based messaging [46].

4. Sensing Techniques and Applications

Two papers on sensing techniques and applications are as follows: (1) "Detection of electronic anklet wearers' groupings throughout telematics monitoring," by Machado et al. [47]; and (2) "Camera coverage estimation based on multistage grid subdivision," by Wang et al. [48].

Machado et al. from Brazil, Spain and Korea in "Detection of electronic anklet wearers' groupings throughout telematics monitoring" proposed sensor data fusion algorithms to analyze the data from anklet positioning devices for tracking convicted individuals. The proposed algorithms can collect and analyze the information of timestamps and locations to estimate the risk assessment. In experimental environments, 10,000 simulated devices generated a set of paths which were obtained from GPS module to evaluate the proposed method, and the response time of the proposed algorithms was evaluated to demo the practicality of the proposed algorithm [47].

Wang et al. from China in "Camera coverage estimation based on multistage grid subdivision" proposed a method based on multistage grid subdivision to efficiently estimate superior camera coverage. This study defined 16 codes of grids, and the grid can be subdivided until each grid can be covered as one of these codes. In experimental environments, the practical data from 15 cameras were collected to evaluate the proposed method, and the results showed that the camera coverage can be estimated by the proposed method with lower time consumption [48].

Author Contributions: C.-H.C. and K.-R.L. edited the special issue, entitled "Applications of Internet of Things", of ISPRS International Journal of Geo-Information. C.-H.C. and K.-R.L. wrote this editorial for the introduction of the special issue.

Funding: This research received no external funding.

Acknowledgments: We would like to thank all authors who submitted their valuable papers to the special issue, entitled "Applications of Internet of Things", of ISPRS International Journal of Geo-Information. Furthermore, we would like to thank all reviewers and the editorial team of ISPRS International Journal of Geo-Information for their great efforts and supports.

Conflicts of Interest: The authors declare no conflict of interest.

References

1. Lin, Y.B.; Lin, Y.W.; Hsiao, C.Y.; Wang, S.Y. Location-based IoT applications on campus: The IoTtalk approach. *Pervasive Mob. Comput.* **2017**, *40*, 660–673. [CrossRef]
2. Chen, C.H.; Lee, C.A.; Lo, C.C. Vehicle localization and velocity estimation based on mobile phone sensing. *IEEE Access* **2016**, *4*, 803–817. [CrossRef]
3. Wu, S.M.; Chen, T.; Wu, Y.J.; Lytras, M. Smart cities in Taiwan: A perspective on big data applications. *Sustainability* **2018**, *10*, 106. [CrossRef]
4. Kung, H.Y.; Kuo, T.H.; Chen, C.H.; Tsai, P.Y. Accuracy analysis mechanism for agriculture data using the ensemble neural network method. *Sustainability* **2016**, *8*, 735. [CrossRef]
5. Ku, H.H. Design of a golf swing injury detection and evaluation open service platform with ontology-oriented clustering case-based reasoning mechanism. *Technol. Health Care* **2015**, *24*, S261–S270. [CrossRef] [PubMed]
6. Marques, G.; Ferreira, C.R.; Pitarma, R. A system based on the Internet of things for real-time particle monitoring in buildings. *Int. J. Environ. Res. Public Health* **2018**, *15*, 821. [CrossRef] [PubMed]
7. Banos, O.; Villalonga, C.; Bang, J.; Hur, T.; Kang, D.; Park, S.; Huynh-The, T.; Le-Ba, V.; Amin, M.B.; Razzaq, M.A.; et al. Human behavior analysis by means of multimodal context mining. *Sensors* **2016**, *16*, 1264. [CrossRef] [PubMed]
8. Lo, C.L.; Chen, C.H.; Kuan, T.S.; Lo, K.R.; Cho, H.J. Fuel consumption estimation system and method with lower cost. *Symmetry* **2017**, *9*, 105. [CrossRef]
9. Lin, J.; Yu, W.; Zhang, N.; Yang, X.; Zhang, H.; Zhao, W. A survey on Internet of things: Architecture, enabling technologies, security and privacy, and applications. *IEEE Internet Things J.* **2017**, *4*, 1125–1142. [CrossRef]
10. Razzaque, M.A.; Milojevic-Jevric, M.; Palade, A.; Clarke, S. Middleware for Internet of things: A survey. *IEEE Internet Things J.* **2016**, *3*, 70–95. [CrossRef]
11. Ruiz-Rosero, J.; Ramirez-Gonzalez, G.; Williams, J.M.; Liu, H.; Khanna, R.; Pisharody, G. Internet of things: A scientometric review. *Symmetry* **2017**, *9*, 301. [CrossRef]
12. Lin, Y.W.; Lin, Y.B.; Hsiao, C.Y.; Wang, Y.Y. IoTtalk-RC: Sensors as universal remote control for aftermarket home appliances. *IEEE Internet Things J.* **2017**, *4*, 1104–1112. [CrossRef]
13. Lin, Y.B.; Wang, S.Y.; Huang, C.C.; Wu, C.M. The SDN approach for the aggregation/disaggregation of sensor data. *Sensors* **2018**, *18*, 2025. [CrossRef] [PubMed]
14. Lin, Y.B.; Cheng, H.Y.; Cheng, Y.H.; Agrawal, P. Implementing automatic location update for follow-me database using VoIP and Bluetooth technologies. *IEEE Trans. Comput.* **2002**, *51*, 1154–1168. [CrossRef]
15. Liou, R.H.; Lin, Y.B.; Sung, Y.C.; Liu, P.C.; Wietfeld, C. Performance of CS fallback for long term evolution mobile network. *IEEE Trans. Veh. Technol.* **2014**, *63*, 3977–3984. [CrossRef]
16. Aijaz, A.; Aghvami, A.H. Cognitive machine-to-machine communications for Internet-of-things: A protocol stack perspective. *IEEE Internet Things J.* **2015**, *2*, 103–112. [CrossRef]
17. Pang, Y.C.; Lin, G.Y.; Wei, H.Y. Context-aware dynamic resource allocation for cellular M2M communications. *IEEE Internet Things J.* **2016**, *3*, 318–326. [CrossRef]
18. Gao, Y.; Qin, Z.; Feng, Z.; Zhang, Q.; Holland, O.; Dohler, M. Scalable and reliable IoT enabled by dynamic spectrum management for M2M in LTE-A. *IEEE Internet Things J.* **2016**, *3*, 1135–1145. [CrossRef]
19. Yin, S.; Bao, J.; Zhang, Y.; Huang, X. M2M security technology of CPS based on Blockchains. *Symmetry* **2017**, *9*, 193. [CrossRef]
20. Jang, U.; Lim, H.; Kim, H. Privacy-Enhancing Security Protocol in LTE Initial Attack. *Symmetry* **2014**, *6*, 1011–1025. [CrossRef]
21. Lin, Y.B.; Wang, L.C.; Chen, W.C. eSES: Enhanced simple energy saving for LTE HeNBs. *IEEE Commun. Lett.* **2017**, *21*, 2520–2523. [CrossRef]

22. Duangmanee, P.; Uthansakul, P. Clock-frequency switching technique for energy saving of microcontroller unit (MCU)-based sensor node. *Energies* **2018**, *11*, 1194. [CrossRef]
23. Chiang, M.; Zhang, T. Fog and IoT: An overview of research opportunities. *IEEE Internet Things J.* **2016**, *3*, 854–864. [CrossRef]
24. Lee, S.; Jeong, T. Cloud-based parameter-driven statistical services and resource allocation in a heterogeneous platform on enterprise environment. *Symmetry* **2016**, *8*, 103. [CrossRef]
25. Lo, C.C.; Chen, C.H.; Cheng, D.Y.; Kung, H.Y. Ubiquitous healthcare service system with context-awareness capability: Design and implementation. *Expert Syst. Appl.* **2011**, *38*, 4416–4436. [CrossRef]
26. Nieuwenhuijsen, M.J.; Donaire-Gonzalez, D.; Foraster, M.; Martinez, D.; Cisneros, A. Using personal sensors to assess the exposome and acute health effects. *Int. J. Environ. Res. Public Health* **2014**, *11*, 7805–7819. [CrossRef] [PubMed]
27. Kung, H.Y.; Chen, C.H.; Ku, H.H. Designing intelligent disaster prediction models and systems for debris-flow disasters in Taiwan. *Expert Syst. Appl.* **2012**, *39*, 5838–5856. [CrossRef]
28. Chen, C.H.; Wu, C.L.; Lo, C.C.; Hwang, F.J. An augmented reality question answering system based on ensemble neural networks. *IEEE Access* **2017**, *5*, 17425–17435. [CrossRef]
29. Zanella, A.; Bui, N.; Castellani, A.; Vangelista, L.; Zorzi, M. Internet of things for smart cities. *IEEE Internet Things J.* **2014**, *1*, 22–32. [CrossRef]
30. Kim, S.; Lim, H. Reinforcement learning based energy management algorithm for smart energy buildings. *Energies* **2018**, *11*, 2010. [CrossRef]
31. Lin, C.Y.; Wang, L.C.; Tsai, K.H. Hybrid Real-Time Matrix Factorization for Implicit Feedback Recommendation Systems. *IEEE Access* **2018**, *6*, 21369–21380. [CrossRef]
32. Lo, C.C.; Kuo, T.H.; Kung, H.Y.; Kao, H.T.; Chen, C.H.; Wu, C.I.; Cheng, D.Y. Mobile merchandise evaluation service using novel information retrieval and image recognition technology. *Comput. Commun.* **2011**, *34*, 120–128. [CrossRef]
33. Chen, C.H. An arrival time prediction method for bus system. *IEEE Internet Things J. Early Access* **2018**. [CrossRef]
34. Lai, W.K.; Kuo, T.H.; Chen, C.H. Vehicle speed estimation and forecasting methods based on cellular floating vehicle data. *Appl. Sci.* **2016**, *6*, 47. [CrossRef]
35. Lin, Y.B.; Lin, Y.W.; Huang, C.M.; Chih, C.Y.; Lin, P. IoTtalk: A management platform for reconfigurable sensor devices. *IEEE Internet Things J.* **2017**, *4*, 1552–1562. [CrossRef]
36. Chen, C.H.; Lin, H.F.; Chang, H.C.; Ho, P.H.; Lo, C.C. An analytical framework of a deployment strategy for cloud computing services: A case study of academic websites. *Math. Probl. Eng.* **2013**, *2013*, 384305. [CrossRef]
37. Lu, N.; Cheng, N.; Zhang, N.; Shen, X.; Mark, J.W. Connected vehicles: Solutions and challenges. *IEEE Internet Things J.* **2014**, *1*, 289–299. [CrossRef]
38. Cui, J.; Wen, J.; Han, S.; Zhong, H. Efficient privacy-preserving scheme for real-time location data in vehicular ad-hoc network. *IEEE Internet Things J. Early Access* **2018**. [CrossRef]
39. Chen, C.H.; Yang, Y.T.; Chang, C.S.; Hsieh, C.M.; Kuan, T.S.; Lo, K.R. The design and implementation of a garbage truck fleet management system. *S. Afr. J. Ind. Eng.* **2016**, *27*, 32–46. [CrossRef]
40. Wu, C.I.; Chen, C.H.; Lin, B.Y.; Lo, C.C. Traffic information estimation methods from handover events. *J. Test. Eval.* **2016**, *44*, 656–664. [CrossRef]
41. Lai, W.K.; Kuo, T.H. Vehicle positioning and speed estimation based on cellular network signals for urban roads. *ISPRS Int. J. Geo-Inf.* **2016**, *5*, 181. [CrossRef]
42. Zhang, Y.; Ye, N.; Wang, R.; Malekian, R. A method for traffic congestion clustering judgment based on grey relational analysis. *ISPRS Int. J. Geo-Inf.* **2016**, *5*, 71. [CrossRef]
43. Ishikawa, T.; Fujinami, K. Smartphone-based pedestrian's avoidance behavior recognition towards opportunistic road anomaly detection. *ISPRS Int. J. Geo-Inf.* **2016**, *5*, 182. [CrossRef]
44. Chen, C.H.; Lin, J.H.; Kuan, T.S.; Lo, K.R. A high-efficiency method of mobile positioning based on commercial vehicle operation data. *ISPRS Int. J. Geo-Inf.* **2016**, *5*, 82. [CrossRef]
45. Wang, H.; Huang, H.; Qin, Y.; Wang, Y.; Wu, M. Efficient location privacy-preserving k-anonymity method based on the credible chain. *ISPRS Int. J. Geo-Inf.* **2017**, *6*, 163. [CrossRef]
46. Gon Jo, H.; Son, T.Y.; Jeong, S.Y.; Kang, S.J. Proximity-based asynchronous messaging platform for location-based Internet of things service. *ISPRS Int. J. Geo-Inf.* **2016**, *5*, 116. [CrossRef]

47. Machado, P.L.; de Sousa, R.T.; de Oliveira Albuquerque, R.; Villalba, L.J.G.; Kim, T.H. Detection of electronic anklet wearers' groupings throughout telematics monitoring. *ISPRS Int. J. Geo-Inf.* **2017**, *6*, 31. [CrossRef]
48. Wang, M.; Liu, X.; Zhang, Y.; Wang, Z. Camera coverage estimation based on multistage grid subdivision. *ISPRS Int. J. Geo-Inf.* **2017**, *6*, 110. [CrossRef]

International Journal of
Geo-Information

Article

Vehicle Positioning and Speed Estimation Based on Cellular Network Signals for Urban Roads

Wei-Kuang Lai and Ting-Huan Kuo *

Department of Computer Science and Engineering, National Sun Yat Sen University, Kaohsiung 804, Taiwan;
wklai@cse.nsysu.edu.tw
* Correspondence: daphnekuo0408@gmail.com; Tel.: +886-7-5252365

Academic Editors: Chi-Hua Chen, Kuen-Rong Lo and Wolfgang Kainz
Received: 20 June 2016; Accepted: 28 September 2016; Published: 2 October 2016

Abstract: In recent years, cellular floating vehicle data (CFVD) has been a popular traffic information estimation technique to analyze cellular network data and to provide real-time traffic information with higher coverage and lower cost. Therefore, this study proposes vehicle positioning and speed estimation methods to capture CFVD and to track mobile stations (MS) for intelligent transportation systems (ITS). Three features of CFVD, which include the IDs, sequence, and cell dwell time of connected cells from the signals of MS communication, are extracted and analyzed. The feature of sequence can be used to judge urban road direction, and the feature of cell dwell time can be applied to discriminate proximal urban roads. The experiment results show the accuracy of the proposed vehicle positioning method, which is 100% better than other popular machine learning methods (e.g., naive Bayes classification, decision tree, support vector machine, and back-propagation neural network). Furthermore, the accuracy of the proposed method with all features (i.e., the IDs, sequence, and cell dwell time of connected cells) is 83.81% for speed estimation. Therefore, the proposed methods based on CFVD are suitable for detecting the status of urban road traffic.

Keywords: intelligent transportation system; cellular networks; vehicle positioning; speed estimation; machine learning

---

1. Introduction

In the last few years, a technical explosion has revolutionized and supported transportation management and control for intelligent transportation systems (ITS). ITS can estimate and obtain traffic information (e.g., traffic flow, traffic density, and vehicle speed) to road users and managers for the improvement of service levels of the road network. The traffic information can be collected and estimated by three approaches, which include: (1) vehicle detection (VD) [1–3]; (2) global positioning system (GPS)-equipped probe car reporting [4–7]; and (3) cellular floating vehicle data (CFVD) [8]. However, vehicle data (VD) has high establishment and maintenance costs. GPS-equipped probe car reporting has a low accuracy rate when the penetration rate of GPS-equipped probe cars is too low. The CFVD can be obtained from mobile phones, which have high penetration in many countries [9], and some studies pointed that CFVD could be used to estimate traffic status with high accuracy [10–27]. Collecting traffic information using CFVD is economic and low cost.

For traffic information estimation based on CFVD, some studies proposed methods to analyze the signals of received signal strength indications (RSSIs), handoffs (HOs), call arrivals (CAs), normal location updates (NLUs), periodical location updates (PLUs), routing area updates (RAUs), and tracking area updates (TAUs). These studies illustrated that higher accuracies of traffic information estimation were performed by using CFVD for highways [10–27]. However, these studies assumed that vehicles can be tracked to the correct route, but the determination of the correct route driven by the user of a mobile station (MS) is difficult and has not been investigated, especially for urban roads.

Therefore, this study proposes a vehicle positioning method to capture CFVD and to track MSs for ITS. Three features of CFVD, which include the IDs, sequence, and cell dwell time of connected cells from the signals of MS communications, are extracted and analyzed. The feature of sequence can be used to judge urban road direction, and the feature of cell dwell time can be applied to discriminate proximal urban roads. Furthermore, this study proposes a vehicle speed estimation method to analyze these three features of CFVD (e.g., IDs, sequence, and cell dwell time of connected cells) for obtaining the real-time estimated vehicle speed.

The rest of this study is organized as follows: the literature reviews of cellular network architecture, CFVD, and traffic information estimation are presented in Section 2; Section 3 proposes a vehicle positioning method based on CFVD to analyze the signals of a mobile phone in a car which is driven on urban roads; a speed estimation method is proposed to measure the speed of the mobile phone in a car according to CFVD in Section 4; the experimental results and discussions are illustrated in Section 5; and Section 6 gives conclusions and discusses future work.

2. Research Background and Related Work

In this section, three subsections, which include cellular networks, CFVD, and traffic information estimation, are discussed for the estimation of traffic information based on CFVD.

2.1. Cellular Networks

This subsection describes the signals and interfaces of cellular networks, which include Global System for Mobile Communications (GSM), General Packet Radio Service (GPRS), Universal Mobile Telecommunications System (UMTS), and Long-Term Evolution (LTE). For circuit-switching networks, MSs can perform the signals of HOs, CAs, NLUs, and PLUs through the A-interface in GSM and through the IuCS-interface in UMTS. For packet-switching networks, MSs can obtain the signals of RAUs through the Gb-interface in GPRS and through the IuPS-interface in UMTS, and the signals of TAUs can be transmitted between MSs and the core network through the S1-MME-interface in LTE [10–27]. Therefore, a network monitor system can be implemented to capture the cellular network signals via the A-interface, the IuCS-interface, the Gb-interface, the IuPS-interface, and the S1-MME-interface for CFVD.

2.2. CFVD

In recent years, CFVD has been analyzed to estimate traffic flow, traffic density, and vehicle speed in some studies. For instance, the signals of HOs from GSM and UMTS could be used to analyze the cell dwell time in cells and to estimate vehicle speed and travel time [8,11,12,16,25,26,28]. Figure 1 shows a case study of CFVD for highway and urban roads. One highway (i.e., Highway 1) and four urban roads (i.e., Urban Road 1, Urban Road 2, Urban Road 3, and Urban Road 4) are covered by three cells (i.e., Cell 1, Cell 2, and Cell 3). When a MS performs a call and moves from Cell 1 to Cell 2, a HO signal is generated and recorded. Moreover, the MS keeps moving from Cell 2 to Cell 3, another HO signal is also generated and recorded. These two HO signals can be analyzed to obtain the cell dwell time of Cell 2. Then the vehicle speed and travel time of Highway 1 can be estimated in accordance with the cell dwell time [8,11,12,16,25,26,28].

Although the previous studies provided high accuracies of traffic information estimation, they focused on highways and assumed that vehicles can be tracked to the correct route. In practical environments, a cell usually covers only one highway, and a cell may cover several urban roads. For instance, Cell 1 covers Highway 1, Urban Road 1, and Urban Road 2. Therefore, the determination of the correct route driven by the MS user is difficult, especially for urban roads.

Some studies proposed a route classification method based on vehicular mobility patterns [12,29,30]. The route classification method recorded the list of cells which covered a same road. For example, the list of cells for Urban Road 1 in Figure 1 is {Cell 1, Cell 2, and Cell 3}. The method could estimate the similarity of the cell list of a route and the list of connected cells of a MS for determining the route which is driven by the MS user [12,29,30]. However, the previous method could not determine the road direction, and the proximal urban roads might lead to lower accuracy of route classification.

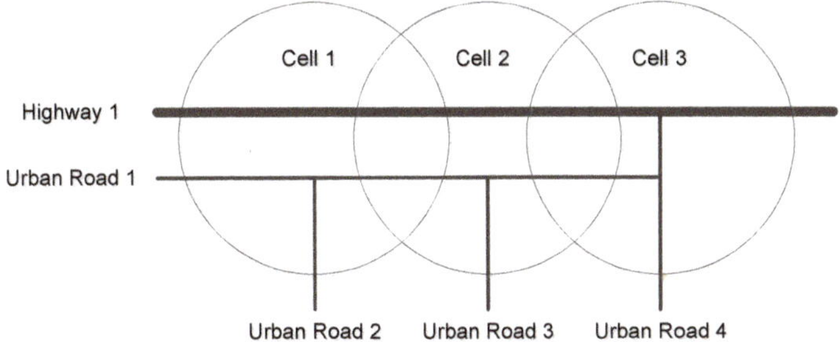

Figure 1. The case study of CFVD for highway and urban roads.

2.3. Traffic Information Estimation

For traffic information estimation, the amount of HOs and NLUs could be collected and analyzed for traffic flow estimation [8,10,14,17], and the amount of CAs and PLUs could be retrieved and used for traffic density estimation [8,10,14,15]. Then the vehicle speed can be estimated in accordance with the estimated traffic flow and the estimated traffic density. Furthermore, some studies proposed mobile positioning methods to measure and analyze RSSIs between the MS and base stations (BSs) to determine the location of the MS [20–23]. The time difference and the distance between two locations of the same MS can be measured for vehicle speed estimation and travel time estimation. The estimated traffic information-based CFVD can be referred and analyzed to develop traffic control strategies for governments.

3. Vehicle Positioning Method

A vehicle positioning method is proposed to collect and analyze CFVD (e.g., the IDs, sequence, and cell dwell time of connected cells) from the signals of MS communications (e.g., call arrivals and handoffs) for determining urban road segments which are driven by MS users in their cars. For instance, Figure 2 shows a case study of an urban road network and cell coverage. There are five cells (i.e., $Cell_1$ to $Cell_5$) and three urban road segments (i.e., $Road_1$ to $Road_3$) in this case. When the MS moves and performs handoff signals, the road segments which are driven by the MS user in their car can be tracked according to the IDs, sequence, and cell dwell time of connected cells. In this case, $Cell_5$, $Cell_4$, $Cell_3$, and $Cell_2$ may be connected by a MS when the MS moves through $Road_1$ to $Road_2$; $Cell_5$, $Cell_4$, $Cell_3$, and $Cell_1$ may be connected by a MS when the MS moves through $Road_1$ to $Road_3$.

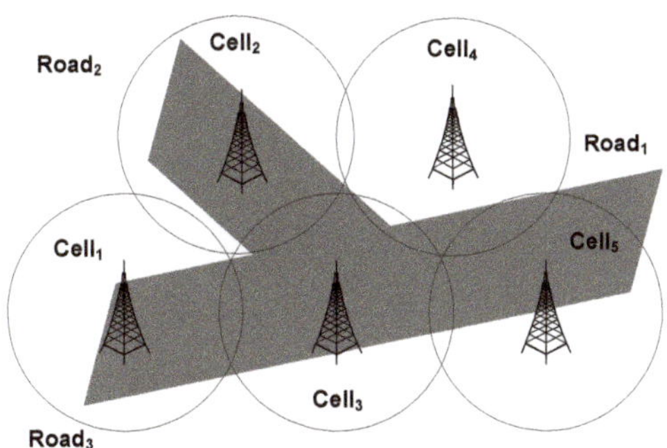

Figure 2. The case study of an urban road network and cell coverage.

Therefore, the proposed vehicle positioning method is designed to analyze CFVD and to apply the *k*-nearest neighbor algorithm (kNN) for determining the location of the vehicle. This method includes four steps (shown in Figure 3) which include: (1) collecting connection and handoff signals from cellular networks; (2) analyzing cell ID, sequence, and cell dwell time of connected cells; (3) retrieving k_1 similar records from a historical dataset; and (4) determining the location of the vehicle. The details of each step are presented in following subsections.

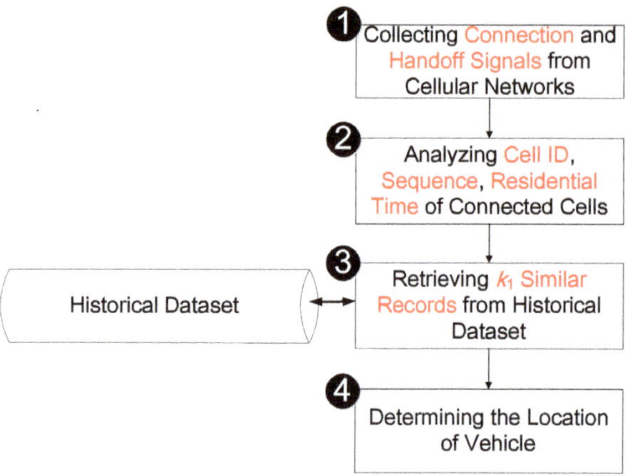

Figure 3. The steps of vehicle positioning method.

3.1. Collecting Connection and Handoff Signals from Cellular Networks

Step 1 captures and collects the cell IDs and timestamps from cellular network signals (e.g., call arrivals and handoffs) which are obtained by MS and core networks via A and IuCS interfaces. This study applies an international mobile subscriber identity (IMSI) as the ID of the MS for tracking each MS. For instance, a call was performed by $IMSI_1$ at PM 16:08:02 on 18 May 2016, and the cellular network signals during this call were collected and showed in Table 1. When this MS moved from $Cell_1$ to $Cell_2$, a handoff procedure was performed at PM 16:10:35. However, cell oscillation might occur between 16:10:35 and 16:11:07. Then, the MS kept moving and entered the coverage of $Cell_3$, and a handoff signal was generated at PM 16:15:58. Finally, a call complete procedure was performed at 16:18:39. These signals can be captured and used as CFVD for vehicle positioning and speed estimation.

Table 1. The cellular network signals during a call performed by $IMSI_1$ on 18 May 2016.

Record ID	Mobile Station ID	Time	Cell ID	Signals
1	$IMSI_1$	18 May 2016 16:08:02	$Cell_1$	Call Arrival
2	$IMSI_1$	18 May 2016 16:10:35	$Cell_2$	Handoff
3	$IMSI_1$	18 May 2016 16:10:46	$Cell_1$	Handoff
4	$IMSI_1$	18 May 2016 16:11:07	$Cell_2$	Handoff
5	$IMSI_1$	18 May 2016 16:15:58	$Cell_3$	Handoff
6	$IMSI_1$	18 May 2016 16:18:39	$Cell_3$	Call Complete

3.2. Analyzing Cell ID, Sequence, and Cell Dwell Time of Connected Cells

Step 2 can analyze the records (i.e., cell IDs and timestamps) from Step 1 and extract three features, which include the cell IDs, sequence, and cell dwell time of connected cells. This study assumes that *n* cells are available in experimental environments. The extraction processes of each feature are illustrated in the following subsections.

3.2.1. Cell ID

For the feature analysis of cell ID, this study sets the value of $Cell_i$ (c_i) as 1 if $Cell_i$ is connected during a call, but otherwise the value of cell is 0. The feature of cell ID, which can be presented as a vector space model (C), is defined in Equation (1). For example, $Cell_1$, $Cell_2$, and $Cell_3$ are connected by $IMSI_1$ in Table 1, so the values of c_1, c_2, and c_3 are 1 (shown in Equation (2)).

$$C = \{c_1, c_2, c_3, c_4, ..., c_n\}, \text{where } c_i = \begin{cases} 1, \text{if } Cell_i \text{ is connected during a call} \\ 0, \text{otherwise} \end{cases} \tag{1}$$

$$C = \{1, 1, 1, 0, ..., 0\} \tag{2}$$

3.2.2. Sequence

For the judgment of urban road direction, the handoff sequence is an important feature, so this study analyzes the sequence of connected cells for determining the road segment driven by a MS user. When $Cell_i$ is firstly connected, the value of $Cell_i$ (o_i) is given with a higher weight value. Then the feature of sequence which can be presented as a vector space model (O) is defined in Equation (3). Furthermore, this study only considers the first x connected cells, and a vector set of weight values (A) for the feature of sequence is defined in Equation (4). For instance, this study set the value of x as 3, and Equation (5) is adopted to set the values of A (i.e., $a_1 = 1$; $a_2 = 0.5$; $a_3 = 0.25$). In the case of $IMSI_1$ in Table 1, $Cell_1$ is firstly connected, so the value of $Cell_1$ (o_1) is given as 1 (i.e., a_1). Then $Cell_2$ is secondly connected, and the value of $Cell_2$ (o_2) is adopted as 0.5 (i.e., a_2). Finally, this study set the value of $Cell_3$ (o_3) as 0.25 (i.e., a_3) and the values of other cells as 0 (shown in Equation (6)).

$$O = \{o_1, o_2, o_3, o_4, ..., o_n\}, \text{where } o_i = \text{the corresponding weight value of } Cell_i \tag{3}$$

$$A = \{a_1, a_2, ..., a_x\} \tag{4}$$

$$A = \{1, 0.5, 0.25\} \tag{5}$$

$$O = \{1, 0.5, 0.25, 0, ..., 0\} \tag{6}$$

3.2.3. Cell Dwell Time

For the discrimination of proximal urban roads, the cell dwell time is an important feature, so this study analyzes the cell dwell time of each connected cell during the same call. However, cell oscillation may occur, especially in a city. Therefore, the total cell dwell time of each cell is considered and summarized. Then, the feature of cell dwell time, which can be presented as a vector space model (T), is defined in Equation (7). Moreover, this study only considers the first y cells with longer cell dwell time, and a vector set of weight values (B) for the feature of cell dwell time is defined in Equation (8). For example, cell oscillation might occur between 16:10:35 and 16:11:07 in Table 1. Therefore, the total cell dwell time of $Cell_1$ is 174 s (i.e., 174 = 153 + 21), and the total cell dwell time of $Cell_2$ is 302 s (i.e., 302 = 11 + 291). Then, the cell dwell time of $Cell_3$ is 161 s. In this study, the value of y is adopted as 3, and Equation (9) is adopted to set the values of B (i.e., $b_1 = 1$; $b_2 = 0.5$; $b_3 = 0.25$). The cell dwell time of $Cell_2$ is the longest in the case of Table 1, so the value of $Cell_2$ (t_2) is given as 1 (i.e., b_1). Then, the values of $Cell_3$ (t_3) and $Cell_1$ (t_1) are adopted as 0.5 (i.e., b_2) and 0.25 (i.e., b_3), respectively. Finally, this study sets the values of other cells as 0 (shown in Equation (10)).

$$T = \{t_1, t_2, t_3, t_4, ..., t_n\}, \text{where } o_i = \text{the corresponding weight value of } Cell_i \tag{7}$$

$$B = \{b_1, b_2, ..., b_x\} \tag{8}$$

$$B = \{1, 0.5, 0.25\} \tag{9}$$

$$T = \{0.25, 1, 0.5, 0, ..., 0\} \tag{10}$$

3.2.4. Combination

This study considers the features of cell ID, sequence, and cell dwell time simultaneously and combines vector space models of C, O, and T into the vector set of R (shown in Equation (11)). For instance, the records of IMSI1 can be modeled in Equation (12):

$$R = \{C, O, T\} = \{c_1, c_2, c_3, c_4, ..., c_n, o_1, o_2, o_3, o_4, ..., o_n, t_1, t_2, t_3, t_4, ..., t_n\} \tag{11}$$

$$R = \{1, 1, 1, 0, ..., 0, 1, 0.5, 0.25, 0, ..., 0, 0.25, 1, 0.5, 0, ..., 0\} \tag{12}$$

3.3. Retrieving k_1 Similar Records from a Historical Dataset

In this study, m calls are transformed in accordance with Equation (11) and stored in a historical database. These m records are defined as historical dataset H (shown in Equation (13)). Furthermore, the driven road segment of each historical record is labeled in the database. When a new call is performed and completed, the vector set of this call (r) (shown in Equation (14)) is transformed according to Equation (11) and compared with each record in historical dataset H by Equation (15). Then the most similar historical record with the distance g_1 can be retrieved in accordance with Equation (16), and Step 3 retrieves k_1 similar records from the historical dataset for vehicle positioning.

$$\begin{aligned}
H &= \{h_1, h_2, ..., h_m\} \\
\text{where } h_i &= \{C_i, O_i, T_i\} \\
&= \{c_{i,1}, c_{i,2}, c_{i,3}, c_{i,4}, ..., c_{i,n}, o_{i,1}, o_{i,2}, o_{i,3}, o_{i,4}, ..., o_{i,n}, t_{i,1}, t_{i,2}, t_{i,3}, t_{i,4}, ..., t_{i,n}\}
\end{aligned} \tag{13}$$

$$r = \{C, O, T\} = \{c_1, c_2, c_3, c_4, ..., c_n, o_1, o_2, o_3, o_4, ..., o_n, t_1, t_2, t_3, t_4, ..., t_n\} \tag{14}$$

$$
\begin{aligned}
d(r, h_i) &= \sqrt{\left(\begin{bmatrix} c_1 - c_{i,1} & \cdots & c_n - c_{i,n} \end{bmatrix} \begin{bmatrix} c_1 - c_{i,1} \\ \vdots \\ c_n - c_{i,n} \end{bmatrix} \begin{bmatrix} o_1 - o_{i,1} & \cdots & o_n - o_{i,n} \end{bmatrix} \begin{bmatrix} o_1 - o_{i,1} \\ \vdots \\ o_n - o_{i,n} \end{bmatrix} \begin{bmatrix} t_1 - t_{i,1} & \cdots & c_n - c_{i,n} \end{bmatrix} \begin{bmatrix} t_1 - t_{i,1} \\ \vdots \\ t_n - t_{i,n} \end{bmatrix}\right) \begin{bmatrix} 1 \\ 1 \\ 1 \end{bmatrix}} \\
&= \sqrt{\sum_{j=1}^{n} (c_j - c_{i,j})^2 + (o_j - o_{i,j})^2 + (t_j - t_{i,j})^2}
\end{aligned} \tag{15}$$

$$g_1 = \min_{1 \leq i \leq m} d(r, h_i) \tag{16}$$

3.4. Determining the Location of a Vehicle

For the determination of vehicle location, Step 4 applies a majority rule to analyze the k_1 similar records, which include the corresponding driven road segment from Step 3. For instance, a case study of a historical dataset and a new record is given in Table 2. There are five cells (i.e., $n = 5$) and six historical records (i.e., $m = 6$), and the value of k_1 is adopted as 3 in this case. Equation (15) is used to calculate the distance between dataset r (i.e., a new record) and each historical record. The result shows that the k_1 similar records are h_1, h_2, and h_4, so Road$_1$ is supported by two records (i.e., h_1 and h_2). Therefore, the driven road segment of this new record is determined as Road$_1$.

Table 2. A case study of historical dataset and a new record.

Record	Road ID	Speed (km/h)	c_1	c_2	c_3	c_4	c_5	o_1	o_2	o_3	o_4	o_5	t_1	t_2	t_3	t_4	t_5
h_1	Road$_1$	60	1	1	1	0	0	1	0.5	0.25	0	0	0.25	1	0.5	0	0
h_2	Road$_1$	58	1	1	1	0	0	1	0.5	0.25	0	0	0.25	0.5	1	0	0
h_3	Road$_1$	40	1	1	1	0	0	0.5	1	0.25	0	0	1	0.5	0.25	0	0
h_4	Road$_2$	59	1	1	1	0	0	0.25	0.5	1	0	0	0.25	1	0.5	0	0
h_5	Road$_2$	50	0	0	1	1	1	0	0	0.5	1	0.25	0	0	1	0.5	0.25
h_6	Road$_2$	53	0	0	1	1	1	0	0	0.25	1	0.5	0	0	0.5	1	0.25
r	?	?	1	1	1	0	0	1	0.5	0.25	0	0	0.25	1	0.5	0	0

4. Speed Estimation Method

This study proposes a method and applies the *k*-nearest neighbor algorithm to extract the features of CFVD (e.g., the IDs, sequence, and cell dwell time of connected cells) and to estimate vehicle speed. The proposed method includes four steps (shown in Figure 4) which include: (1) determining the location of a vehicle; (2) analyzing cell ID, sequence, and cell dwell time of connected cells; (3) retrieving k_2 similar records with the same road segment from historical dataset; and (4) estimating the speed of a vehicle. The details of each step are presented in following subsections.

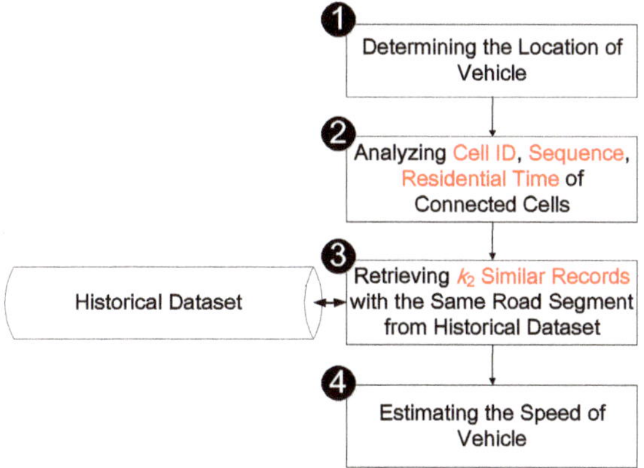

Figure 4. The steps of speed estimation method.

4.1. Determining the Location of Vehicle

Step 1 determines the driven road segment of the MS in accordance with CFVD and the proposed vehicle positioning method in Section 3. This study only considers and analyzes the historical records with the same road segment to estimate vehicle speed. For example, when a new record is determined as $Road_l$, the historical records with $Road_l$ are considered in the following steps.

4.2. Analyzing Cell ID, Sequence, and Cell Dwell Time of Connected Cells

Step 2 adopts Equations (1), (3) and (7) to extract the features of historical records and new records which include the IDs, sequence, and cell dwell time of connected cells. Each record can be transformed as a vector space model (shown in Equation (11)). Historical records are presented as a vector set H, and a new record is presented as a vector set r in accordance with Equations (13) and (14).

4.3. Retrieving k_2 Similar Records with the Same Road from Historical Dataset

Step 3 retrieves k_2 similar records with the same road segment from a historical dataset according to Equation (15). Furthermore, the vehicle speed of each historical record is labeled in a database. For instance, in the case of Table 2, the new record r is determined as $Road_1$, so three historical records (i.e., h_1, h_2, and h_3) are considered to be analyzed for vehicle speed estimation. If the value of k_2 is adopted as 2 in this case, the records h_1 and h_2 are retrieved as the k_2 similar records.

4.4. Estimating the Speed of a Vehicle

Step 4 applies a weighted mean method to analyze the k_2 similar records for vehicle speed estimation. In this study, new record r is determined as $Road_l$, and the distance between this record and the more similar record with vehicle speed v_1 is defined as p_1 in Equation (17). Moreover, the distance between this record and the j-th most similar record with vehicle speed v_j is defined as p_j. Then the vehicle speed of this record is estimated as u by Equation (18). For example, the k_2 similar records are h_1 and h_2 in Table 2 when the value of k_2 is 2. The value of $d(r, h_1)$ is 0 (i.e., $p_1 = 0$), and the value of $d(r, h_2)$ is about 0.707 (i.e., $p_2 = 0$). Then, Equation (18) is adopted to estimate the vehicle speed of the new record r as 60 km/h (shown in Equation (19)).

$$p_1 = \min d\,(r, h_i) \text{ where the driven road segment of } h_i \text{ is } Road_l \tag{17}$$

$$u = \frac{\begin{bmatrix} \dfrac{p_{k_2}-p_1}{p_{k_2}-p_1} & \dfrac{p_{k_2}-p_2}{p_{k_2}-p_1} & \cdots & \dfrac{p_{k_2}-p_{k_2}}{p_{k_2}-p_1} \end{bmatrix} \begin{bmatrix} v_1 \\ v_2 \\ \vdots \\ v_{k_2} \end{bmatrix}}{\begin{bmatrix} \dfrac{p_{k_2}-p_1}{p_{k_2}-p_1} & \dfrac{p_{k_2}-p_2}{p_{k_2}-p_1} & \cdots & \dfrac{p_{k_2}-p_{k_2}}{p_{k_2}-p_1} \end{bmatrix} \begin{bmatrix} 1 \\ 1 \\ \vdots \\ 1 \end{bmatrix}}$$

$$= \frac{\begin{bmatrix} \omega_1 & \omega_2 & \cdots & \omega_{k_2} \end{bmatrix} \begin{bmatrix} v_1 \\ v_2 \\ \vdots \\ v_{k_2} \end{bmatrix}}{\begin{bmatrix} \omega_1 & \omega_2 & \cdots & \omega_{k_2} \end{bmatrix} \begin{bmatrix} 1 \\ 1 \\ \vdots \\ 1 \end{bmatrix}} \text{ where } \omega_i = \frac{p_{k_2}-p_i}{p_{k_2}-p_1}$$

$$= \frac{\sum_{i=1}^{k_2}(\omega_i \times v_i)}{\sum_{i=1}^{k_2}\omega_i} \text{ where } \omega_i = \frac{p_{k_2}-p_i}{p_{k_2}-p_1} \tag{18}$$

$$\begin{aligned} u &= \frac{\omega_1 \times 60 + \omega_2 \times 58}{\omega_1 + \omega_2} \text{ where } \omega_1 = \frac{0.707-0}{0.707-0} = 1 \text{ and } \omega_i = \frac{0.707-0.707}{0.707-0} = 0 \\ &= 60 \end{aligned} \tag{19}$$

5. Experimental Results and Discussions

The collection of CFVD and the information of urban road networks are presented in Section 5.1. The collected CFVD is used to evaluate the proposed vehicle positioning method and speed estimation method in Sections 5.2 and 5.3, respectively.

5.1. Experimental Environment

In experimental environments, a MS (e.g., HTC (Taoyuan, Taiwan) M8 running the Android 2.2.2platform) is carried in a car to perform call procedures when the car is driven on urban roads, and the cellular network signals of these calls can be captured for the collection of CFVD. Six urban road segments in Kaohsiung and Pingtung in Taiwan (shown in Figure 5) are driven in 27 runs. There are 64 different base stations (BSs) (i.e., $n = 64$) detected on these road segments in Taiwan.

For the evaluations of the vehicle positioning method and speed estimation method, some popular machine learning methods (e.g., kNN, naive Bayes classification (NB), decision tree (DT), support vector machine (SVM), and back-propagation neural network (BPNN) [31,32]), are implemented and compared by using the R language [33,34] and Rstudio [35] to analyze collected CFVD in experiments. This study uses the packages of class [36], e1071 [37], party [38], and neuralnet [39] to implement kNN, NB, DT, SVM, and BPNN algorithms, respectively. Furthermore, the k-fold cross-validation method [31,32] is used to analyze each test run. In the i-th iteration, the data of the i-th run is selected as the test corpus, and the other test runs are collectively used to be training data for performance analyses.

Figure 5. The urban road segments in the experimental environment.

5.2. The Evaluation of Vehicle Positioning Method

For the evaluation of the vehicle positioning method, this study considers different features and machine learning methods to analyze CFVD. Considering cell ID and kNN first; it can be observed that its performance of vehicle positioning is 51.85% (shown in Table 3). The cause of several errors is direction misjudgment when only the feature of cell ID is considered. Then, the features of cell ID and sequence are considered for the judgment of urban road direction, and the results show that the accuracy of the vehicle positioning method is improved to 92.59%. However, some proximal urban roads cannot be discriminated by using the features of cell ID and sequence. Finally, this study analyzes all features (i.e., the IDs, sequence, and cell dwell time of connected cells) to determine the driven road segment of the MS user, and the accuracy can be improved to 100%. Therefore, the feature of cell dwell time can support for the discrimination of proximal urban roads.

Table 3. The comparisons of the proposed method with different features for vehicle positioning.

Feature	Accuracy
Cell ID (Previous method [12,29])	51.85%
Cell ID and sequence	92.59%
Cell ID and cell dwell time	88.89%
Cell ID, sequence, and cell dwell time	100%

For the comparisons of different machine learning methods, all features are considered and analyzed to determine the driven road segment. Four factors, which include precision, recall, F_1 − measure (shown in Equation (20)), and accuracy are used to evaluate the performance of each method. Table 4 shows that the performance of the proposed method is higher than other methods.

$$F_1 - measure = \frac{2}{\frac{1}{Precision} + \frac{1}{Recall}}$$

(20)

Table 4. The comparisons of different machine learning methods with all features for vehicle positioning.

Method	Precision	Recall	F_1-Measure	Accuracy
Naive Bayes classification	91.90%	88.33%	90.08%	88.89%
Decision tree	11.67%	20.00%	14.74%	22.22%
Support vector machine	27.78%	50.00%	35.71%	55.56%
Back-propagation neural network	65.83%	56.67%	60.91%	59.26%
The proposed method	100%	100%	100%	100%

For the comparisons of different parameters, this study designs five cases which include {1, 0, 0}, {1, 0.5, 0}, {1, 1, 1}, {1, 0.67, 0.33}, and {1, 0.5, 0.25} for the values of A and B. Furthermore, Euclidean distance, Minkowski distance, and Mahalanobis distance are considered for the proposed method. The experimental results of these cases (in Table 5) indicated that the parameters A and B can be adapted as {1, 0.5, 0.25} to obtain a higher accuracy of vehicle positioning.

Table 5. The comparisons of different parameters for vehicle positioning.

A	B	Distance Method	F_1-Measure	Accuracy
{1, 0, 0}	{1, 0, 0}	Euclidean	87.50%	88.89%
{1, 0.5, 0}	{1, 0.5, 0}	Euclidean	91.67%	92.59%
{1, 1, 1}	{1, 1, 1}	Euclidean	89.17%	88.89%
{1, 0.67, 0.33}	{1, 0.67, 0.33}	Euclidean	96.67%	96.30%
{1, 0.5, 0.25}	{1, 0.5, 0.25}	Euclidean	100.00%	100.00%
{1, 0, 0}	{1, 0, 0}	Minkowski	87.50%	88.89%
{1, 0.5, 0}	{1, 0.5, 0}	Minkowski	91.67%	92.59%
{1, 1, 1}	{1, 1, 1}	Minkowski	92.50%	92.59%
{1, 0.67, 0.33}	{1, 0.67, 0.33}	Minkowski	96.67%	96.30%
{1, 0.5, 0.25}	{1, 0.5, 0.25}	Minkowski	100.00%	100.00%
{1, 0, 0}	{1, 0, 0}	Mahalanobis	87.50%	88.89%
{1, 0.5, 0}	{1, 0.5, 0}	Mahalanobis	91.67%	92.59%
{1, 1, 1}	{1, 1, 1}	Mahalanobis	89.17%	88.89%
{1, 0.67, 0.33}	{1, 0.67, 0.33}	Mahalanobis	96.67%	96.30%
{1, 0.5, 0.25}	{1, 0.5, 0.25}	Mahalanobis	100.00%	100.00%

5.3. The Evaluation of the Speed Estimation Method

For the evaluation of the speed estimation method, this study considers different features after determining the driven road segment of the MS user. Table 6 shows the results of the proposed speed estimation method with different features. These experimental results indicate that cell dwell time is the most important feature, and the accuracy of vehicle estimation with all features can be improved to 83.81%. Therefore, the proposed method based on CFVD is suitable for detecting the status of urban road traffic.

Table 6. The comparisons of the proposed method with different features for speed estimation.

Feature	Accuracy
Cell ID and sequence	78.34%
Cell ID and cell dwell time	80.86%
Cell ID, sequence, and cell dwell time	83.81%

6. Conclusions and Future Work

Several studies of CFVD focused on the traffic information estimation for freeways. Furthermore, these studies assumed that the cellular network signals from the moving MSs on roads can be filtered. However, a cell may cover several road segments of urban roads, so the assumption may not be realized on urban roads.

Therefore, this study proposes vehicle positioning and speed estimation methods to capture CFVD and to track MSs for intelligent transportation systems. Three features of CFVD, which include the IDs, sequence, and cell dwell time of connected cells from the signals of MS communications, are extracted and analyzed. The feature of sequence can be used to judge the urban road direction, and the feature of cell dwell time can be applied to discriminate proximal urban roads. The experiment results show that the accuracy of the proposed vehicle positioning method is better than other popular machine learning methods (e.g., NB, DT, SVM, and BPNN). Furthermore, the accuracy of the proposed method with all features (i.e., the IDs, sequence, and cell dwell time of connected cells) is 83.81% for speed estimation.

However, cell oscillation problems may disturb the cell dwell time of each cell and vehicle speed estimation. This study summarizes the total cell dwell time of each cell to solve these problems, but these problems may occur in accordance with some environment factors. Therefore, the environmental factors may be analyzed to filter out cell oscillation in future work.

Acknowledgments: The research is supported by the Ministry of Science and Technology of Taiwan under the grant No. MOST 104-2221-E-110-041.

Author Contributions: Wei-Kuang Lai and Ting-Huan Kuo proposed and designed the vehicle positioning and speed estimation methods based on CFVD. Ting-Huan Kuo performed the proposed methods and reported the experimental results. All of the authors have read and approved this manuscript.

Conflicts of Interest: The authors declare no conflict of interest.

References

1. Jang, J.; Byun, S. Evaluation of traffic data accuracy using Korea detector testbed. *IET Intell. Transp. Syst.* **2011**, *5*, 286–293. [CrossRef]
2. Ramezani, A.; Moshiri, B.; Kian, A.R.; Aarabi, B.N.; Abdulhai, B. Distributed maximum likelihood estimation for flow and speed density prediction in distributed traffic detectors with Gaussian mixture model assumption. *IET Intell. Transp. Syst.* **2012**, *6*, 215–222. [CrossRef]
3. Middleton, D.; Parker, R. *Vehicle Detector Evaluation*; Report No. FHWA/TX-03 /2119-1; Texas Transportation Institute, Texas Department of Transportation: Austin, TX, USA, 2002.
4. Chen, W.J.; Chen, C.H.; Lin, B.Y.; Lo, C.C. A traffic information prediction system based on global position system-equipped probe car reporting. *Adv. Sci. Lett.* **2012**, *16*, 117–124. [CrossRef]
5. Hunter, T.; Herring, R.; Abbeel, P.; Bayen, A. Path and travel time inference from GPS probe vehicle data. In Proceedings of the Neural Information Processing Foundation Conference, Vancouver, BC, Canada, 5–10 December 2009.
6. Cheu, R.L.; Xie, C.; Lee, D.H. Probe vehicle population and sample size for arterial speed estimation. *Comput. Aided Civil Infrastruct. Eng.* **2002**, *17*, 53–60. [CrossRef]
7. Herrera, J.C.; Work, D.B.; Herring, R.; Ban, X.J.; Jacoboson, Q.; Bayen, A.M. Evaluation of traffic data obtained via GPS-enabled mobile phones: The mobile century field experiment. *Transp. Res. Part C Emerg. Technol.* **2010**, *18*, 568–583. [CrossRef]
8. Caceres, N.; Wideberg, J.P.; Benitez, F.G. Review of traffic data estimations extracted from cellular networks. *IET Intell. Transp. Syst.* **2008**, *2*, 179–192. [CrossRef]
9. United Marketing Research. The Investigation Report on the Digital Opportunity about Phone Users. Research, Development and Evaluation Commission, Executive Yuan, 2011. Available online: http://www.rdec.gov.tw/public/Attachment/213014313671.pdf (accessed on 20 June 2016).
10. Lai, W.K.; Kuo, T.H.; Chen, C.H. Vehicle speed estimation and forecasting methods based on cellular floating vehicle data. *Appl. Sci.* **2016**, *6*, 47. [CrossRef]
11. Wu, C.I.; Chen, C.H.; Lin, B.Y.; Lo, C.C. Traffic information estimation methods from handover events. *J. Test. Eval.* **2016**, *44*, 656–664. [CrossRef]
12. Chang, M.F.; Chen, C.H.; Lin, Y.B.; Chia, C.Y. The frequency of CFVD speed report for highway traffic. *Wirel. Commun. Mob. Comput.* **2015**, *15*, 879–888. [CrossRef]
13. Janecek, A.; Valerio, D.; Hummel, K.A.; Ricciato, F.; Hlavacs, H. The cellular network as a sensor: From mobile phone data to real-time road traffic monitoring. *IEEE Trans. Intell. Transp. Syst.* **2015**, *16*, 2551–2572. [CrossRef]

14. Chen, C.H.; Chang, H.C.; Su, C.Y.; Lo, C.C.; Lin, H.F. Traffic speed estimation based on normal location updates and call arrivals from cellular networks. *Simul. Model. Pract. Theory* **2013**, *35*, 26–33. [CrossRef]
15. Chang, H.C.; Chen, C.H.; Lin, B.Y.; Kung, H.Y.; Lo, C.C. Traffic information estimation using periodic location update events. *Int. J. Innov. Comput. Inf. Control* **2013**, *9*, 2031–2041.
16. Maerivoet, S.; Logghe, S. Validation of travel times based on cellular floating vehicle data. In Proceedings of the 6th European Congress and Exhibition on Intelligent Transport Systems and Services, Aalborg, Denmark, 18–20 June 2007.
17. Caceres, N.; Romero, L.M.; Benitez, F.G.; del Castillo, J.M. Traffic flow estimation models using cellular phone data. *IET Intell. Transp. Syst.* **2012**, *13*, 1430–1441. [CrossRef]
18. Valerio, D.; Witek, T.; Ricciato, F.; Pilz, R.; Wiedermann, W. Road traffic estimation from cellular network monitoring: A hands-on investigation. In Proceedings of the IEEE 20th International Symposium on Personal, Indoor and Mobile Radio Communications, Tokyo, Japan, 13–16 September 2009.
19. Valerio, D.; D'Alconzo, A.; Ricciato, F.; Wiedermann, W. Exploiting cellular networks for road traffic estimation: A survey and a research roadmap. In Proceedings of the IEEE 69th Vehicular Technology Conference, Barcelona, Spain, 26–29 April 2009.
20. Chen, C.H.; Lo, C.C.; Lin, H.F. The Analysis of Speed-Reporting Rates from a cellular network based on a fingerprint-positioning algorithm. *S. Afr. J. Ind. Eng.* **2013**, *24*, 98–106. [CrossRef]
21. Chen, C.H.; Lin, B.Y.; Chang, H.C.; Lo, C.C. The novel positioning algorithm based on cloud computing—A case study of intelligent transportation systems. *Inf. Int. Interdiscip. J.* **2012**, *15*, 4519–4524.
22. Cheng, D.Y.; Chen, C.H.; Hsiang, C.H.; Lo, C.C.; Lin, H.F.; Lin, B.Y. The optimal sampling period of a fingerprint positioning algorithm for vehicle speed estimation. *Math. Probl. Eng.* **2013**, *2013*. [CrossRef]
23. Chen, C.H.; Lin, B.Y.; Lin, C.H.; Liu, Y.S.; Lo, C.C. A green positioning algorithm for campus guidance system. *Int. J. Mob. Commun.* **2012**, *10*, 119–131. [CrossRef]
24. Gundlegård, D.; Karlsson, J.M. The smartphone as enabler for road traffic information based on cellular network signaling. In Proceedings of the 16th International IEEE Conference on Intelligent Transportation Systems, Hague, The Netherlands, 6–9 October 2013.
25. Gundlegard, D.; Karlsson, J.M. Handover location accuracy for travel time estimation in GSM and UMTS. *IET Intell. Transp. Syst.* **2009**, *3*, 87–94. [CrossRef]
26. Gundlegard, D.; Karlsson, J.M. Route classification in travel time estimation based on cellular network signaling. In Proceedings of the 12th International IEEE Conference on Intelligent Transportation Systems, St. Louis, MO, USA, 4–7 October 2009.
27. Gundlegard, D.; Karlsson, J.M. Generating road traffic information from cellular networks—New possibilities in UMTS. In Proceedings of the 6th International Conference on ITS Telecommunications, Chengdu, China, 21–23 June 2006.
28. Demissie, M.G.; de Almeida Correia, G.H.; Bento, C. Intelligent road traffic status detection system through cellular networks handover information: An exploratory study. *Transp. Res. Part C Emerg. Technol.* **2013**, *32*, 76–88. [CrossRef]
29. Fiadino, P.; Valerio, D.; Ricciato, F.; Hummel, K.A. Steps towards the extraction of vehicular mobility patterns from 3G signaling data. *Lect. Notes Comput. Sci.* **2012**, *7189*, 66–80.
30. Becker, R.A.; Caceres, R.; Hanson, K.; Loh, J.M.; Urbanek, S.; Varshavsky, A.; Volinsky, C. Route classification using cellular handoff patterns. In Proceedings of the 13th International Conference on Ubiquitous Computing, Beijing, China, 17–21 September 2011.
31. Lai, W.K.; Kuo, T.H. An urban road segment determination method based on cellular floating vehicle data for tracking mobile stations. In Proceedings of the 7th International IEEE Conference on Ubi-Media Computing and Workshops, Ulaanbaatar, Mongolia, 12–14 July 2014.
32. Han, J.; Kamber, M.; Pei, J. *Data Mining: Concepts and Techniques*, 3rd ed.; Morgan Kaufmann Publishers: San Francisco, CA, USA, 2011.
33. Ihaka, R.; Gentleman, R. R: A language for data analysis and graphics. *J. Comput. Graph. Stat.* **1996**, *5*, 299–314.
34. Ripley, B.D. The R project in statistical computing. *MSOR Connect.* **2001**, *1*, 23–25. [CrossRef]
35. Racine, J.S. RStudio: A platform-independent IDE for R and sweave. *J. Appl. Econ.* **2012**, *27*, 167–172. [CrossRef]
36. Ripley, B.; Venables, W. Class: Functions for Classification. The Comprehensive R Archive Network 2015. Available online: https://cran.r-project.org/web/packages/class/index.html (accessed on 20 June 2016).

37. Meyer, D.; Dimitriadou, E.; Hornik, K.; Weingessel, A.; Leisch, F.; Chang, C.C.; Lin, C.C. e1071: Nisc Functions of the Department of Statistics, Probability Theory Group. The Comprehensive R Archive Network 2015. Available online: https://cran.r-project.org/web/packages/e1071/index.html (accessed on 20 June 2016).
38. Hothorn, T.; Hornik, K.; Strobl, C.; Zeileis, A. Party: A Laboratory for Recursive Partytioning. The Comprehensive R Archive Network 2015. Available online: https://cran.r-project.org/web/packages/party/index.html (accessed on 20 June 2016).
39. Fritsch, S.; Guenther, F. Neuralnet: Training of Neural Networks. The Comprehensive R Archive Network 2012. Available online: https://cran.r-project.org/web/packages/neuralnet/index.html (accessed on 20 June 2016).

International Journal of
Geo-Information

Article

A Method for Traffic Congestion Clustering Judgment Based on Grey Relational Analysis

Yingya Zhang [1], Ning Ye [1,2,*,†], Ruchuan Wang [3,†] and Reza Malekian [4,*]

[1] Department of Computer, Nanjing University of Post and Telecommunications, Nanjing 210003, China; 1014041108@njupt.edu.cn
[2] Jiangsu High Technology Research Key Laboratory for Wireless Sensor Networks, Nanjing 210003, China
[3] Key Lab of Broadband Wireless Communication and Sensor Network Technology of Ministry of Education, Nanjing University of Post and Telecommunications, Nanjing 210003, China; wangrc@njupt.edu.cn
[4] Departamento de Ingeniería Informática, Universidad de Santiago de Chile, Av. Ecuador, Santiago 3659, Chile
* Correspondence: yening@njupt.edu.cn (N.Y.); reza.malekian@ieee.org (R.M.);
  Tel.: +86-138-1389-2478 (N.Y.); +27-12-420-4305 (R.M.);
  Fax: +86-025-83492470 (N.Y.); +27-12-420-362-5000 (R.M.)
† These authors contributed equally to this work.

Academic Editors: Chi-Hua Chen, Kuen-Rong Lo and Wolfgang Kainz
Received: 17 January 2016; Accepted: 9 May 2016; Published: 18 May 2016

Abstract: Traffic congestion clustering judgment is a fundamental problem in the study of traffic jam warning. However, it is not satisfactory to judge traffic congestion degrees using only vehicle speed. In this paper, we collect traffic flow information with three properties (traffic flow velocity, traffic flow density and traffic volume) of urban trunk roads, which is used to judge the traffic congestion degree. We first define a grey relational clustering model by leveraging grey relational analysis and rough set theory to mine relationships of multidimensional-attribute information. Then, we propose a grey relational membership degree rank clustering algorithm (GMRC) to discriminant clustering priority and further analyze the urban traffic congestion degree. Our experimental results show that the average accuracy of the GMRC algorithm is 24.9% greater than that of the K-means algorithm and 30.8% greater than that of the Fuzzy C-Means (FCM) algorithm. Furthermore, we find that our method can be more conducive to dynamic traffic warnings.

Keywords: urban traffic; grey relational membership degree; traffic congestion judgment

---

1. Introduction

With the rapid development of urban traffic, urban vehicle surges and the pressure on traffic capacities are increasing sharply. Therefore, traffic problems are becoming serious and bound the development of a city. In China, the conditions of roads and vehicles are quite inconvenient, and traffic congestion has caused substantial social and economic problems. In this case, traffic jams not only waste time, delay work, and reduce efficiency but also cause a substantial waste of fuel, increase the probability of accidents and exacerbate the already serious difficulties facing traffic control and management. Since the 1980s, intelligent transportation systems (ITSs) consisting of integrated computer technology, automatic control technology, communication technology and information processing technology have achieved remarkable results worldwide. In addition, many aspects of ITSs are based on traffic information. Furthermore, traffic information processing has become an important aspect of ITSs [1]. The critical function of an ITS is to manage and control traffic flows and avoid the development of traffic jams. When traffic jams occur, such systems should provide timely and effective solutions and ease traffic pressure. Therefore, clustering and evaluating urban traffic congestion is of great importance and is thus a prerequisite for correctly inspecting traffic congestion. To determine the road congestion degree, different definitions of traffic congestion are formulated. Rothenberg defined the traffic congestion rank as the number of vehicles on the road exceeding the carrying capacity on the general acceptable road service level [2]. Under such a definition, U.S. authorities divide Level of Service (LOS) into six levels from A to F based on the ratio of the actual vehicle flow (volume) and road capacity: V/C. In Virginia, when V/C is less than 0.77, LOS is at the D level, and the traffic situation is considered to be a high-density but stable traffic flow. When LOS is at the E level, traffic begins to deteriorate and results in a serious traffic jam. A method of assessing the traffic congestion level (rank) is by comparing a certain traffic parameter with a threshold. When the parameter is greater than a certain threshold, a traffic jam is considered to have formed. Specifically, the method can detect whether traffic congestion occurs but cannot represent a comprehensive information evaluation method for traffic congestion. Currently, we collect traffic flow information based on three properties for Nanjing urban trunk roads to comprehensively weigh the level of traffic congestion in the same time period. In addition, we judge which road is allowing smooth traffic flows, which is suffering from a light traffic jam, which is suffering from a traffic jam, and which is suffering from a heavy traffic jam state. Figure 1 represents the Nanjing transport network area in its geographical context. Using the above information, we can provide reference values for traffic management.

At present, the method of judging traffic congestion can be divided into three categories: (1) Direct detection method, such as video detection method. This method needs to install too many cameras and the cost is higher. (2) Indirect detection method, which is mainly according to events' influence on traffic flow, used to detect the event's existence. The method has low cost, simple and easy to operate, but has lower detection rate, and high false alarm rate. (3) Based on theory model, design algorithm to judge traffic congestion. This method is being used, and some mature theories have been applied, such as cluster analysis, grey system theory, and rough set theory. Our work concentrates on clustering traffic flow information based on grey relational analysis to judge traffic congestion situations [3].

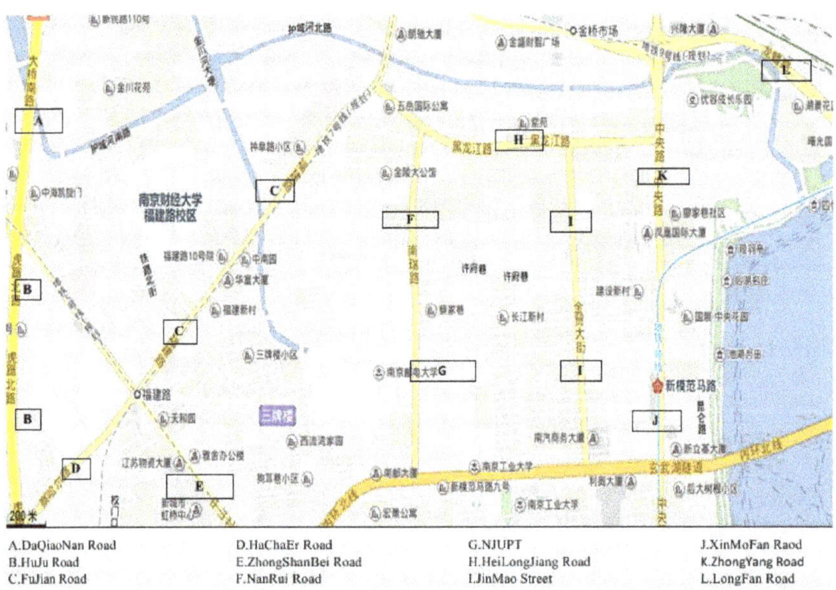

A.DaQiaoNan Road	D.HuChaEr Road	G.NJUPT	J.XinMoFan Raod
B.HuJu Road	E.ZhongShanBei Road	H.HeiLongJiang Road	K.ZhongYang Road
C.FuJian Road	F.NanRui Road	I.JinMao Street	L.LongFan Road

Figure 1. Nanjing transport network area in its geographical context.

The paper is organized as follows. First, we give a brief summary of previous related work in Section 2. Next, we introduce how to build the grey relational clustering model in Section 3.

In Section 4, the grey relational membership degree rank clustering algorithm (GMRC) is described. Then, we illustrate experimental results in Section 5. Finally, we conclude the paper in Section 6.

2. Related Work

2.1. Related Theories

Our work particularly involves grey system theory. Grey system theory was proposed by Professor Julong Deng in 1982 and includes many aspects such as grey generation, grey analysis, grey modeling and grey prediction [4]. This theory has been widely used in image processing, network security and logistics management. In addition, grey relational analysis is not only an important aspect of grey theory but also a type of measurement method for the study of the similarity of data. This analysis also has no strict requirement on sample size, and it is usually used for analysis and comparison of the geometric forms of curves described by several points in the space; the closer to the referenced standard array, the higher the relational degree between the referenced standard and the higher its rank. Next, we will briefly introduce rough set theory, which is a mathematical theory method used to address uncertain, imprecise and fuzzy information. This theory has been applied in machine learning, data mining, decision-making analysis, *etc*. In addition, rough set theory is an important branch of uncertainty calculations, which include other theories such as fuzzy sets, neural networks and evidence theory. In this paper, we analyze urban road traffic information using grey relational clustering and combine the results with rough set theory to establish a decision table system. Finally, we judge the degree of urban traffic congestion.

2.2. Clustering Techniques

Clustering analysis is the subject of active research in several fields such as statistics, pattern recognition, machine learning, and data mining. It aims to partition a large number of data into different subsets or groups so that the requirements of homogeneity and heterogeneity are fulfilled. Homogeneity requires that data in the same cluster should be as similar as possible and heterogeneity means that data in different clusters should be as different as possible.

At present, a number of cluster methods have been widely used, among which a weighted adaptive algorithm based on the equilibrium approach is proposed wherein the grey method is introduced to a spectral clustering algorithm [5] to measure the similarity between the data. The experimental results show that the proposed algorithm can effectively overcome the shortcomings of spectral clustering concerning the sensitivity of parameters. Another clustering algorithm based on entropy that can automatically determine the number of clusters based on the distribution characteristics of the data sample and reduce the user's participation was proposed in [6]. The result is more objective, and the algorithm can find large and small clusters of any shape. The disadvantages of the algorithm are the selection of the initial points and the effects of noise and outliers. However, these shortcomings can be overcome by screening and addressing the original data noise and outliers, excluding false data and improving the reliability and separation of the data to minimize the influence of noise and outliers. The two methods inspire our work such that we consider using an approach to comprehensively evaluate data; therefore, we use grey relational theory, which can effectively process multidimensional attribute data. However, those traditional algorithms are mostly a simple clustering of similar data and do not consider what data attributes have what indicator characteristics. Moreover, clustering results cannot reflect the rank of data that present greater clustering.

2.3. K-Means Algorithm and FCM Algorithm

Currently, the K-means algorithm and Fuzzy C-Means (FCM) algorithm are commonly used to cluster data. We consider the two algorithms comparing with our proposed algorithm in our work. First, our work will provide a brief introduction of the K-means algorithm. The K-means algorithm is a classical clustering algorithm that is widely used in different subject areas. In addition, various improved algorithms have evolved based on the K-means algorithm. The K-means algorithm, however, is an NP-hard optimization problem (Generally speaking, problems that will cost polynomial time to solve and are easy to address are commonly regarded as P problems. Problems that cost super polynomial time to solve are considered as difficult problems and are known as NP-hard problems), which means that many problems cannot obtain global optimal results [7]. The Euclidean distance is typically used as the criterion function of the K-means algorithm, where the distance between two data points with different units is sometimes calculated. The Euclidean distance is the real distance between two points in an m-dimensional space and is mainly determined by the heavily fluctuant elements. Slightly fluctuant elements are often neglected, and the phenomenon is more obvious with increasing ratio of the difference between corresponding elements.

FCM algorithm is a type of fuzzy clustering algorithm based on the objective function. First, we introduce the concept of fuzziness: fuzziness is uncertainty. In addition, we usually say that an object is what is certain. However, the uncertainty indicates the similarity between two objects [8,9]. For example, we regard twenty years of age as a standard of judging whether an individual is young or not young. Therefore, a 21-year-old person is not young according to this division of certainty; however, 21 years of age is very young in our opinion. Thus, at this moment, we can vaguely think of the possibility of a 21-year-old person belonging to young as 90% and that belonging to not young as 10%. Here, 90% and 10% are not probabilities; rather, they are the degree of similarity. Although the FCM algorithm can effectively perform clustering, it does not differentiate the rank to which the clustering belongs.

In view of the above problems, we mine traffic flow information relations with different attributes via grey relational clustering. In addition, we propose the GMRC algorithm to judge the clustering priority. Simultaneously, we combine the K-means and FCM clustering algorithms and contrast the results with those of the GMRC algorithm to evaluate the performance of our algorithm.

3. Grey Relational Clustering Model

In this paper, we suppose that the traffic congestion state of roads is divided into four ranks (not four clusters), namely, smooth, light jam, jam and heavy jam, which is our precondition, where smooth characterizes the best condition, belonging to the first rank, and heavy jam characterizes the worst condition, belonging to the fourth rank. According to the description of the different definitions of traffic congestion, traffic congestion is not only related to certain parameters [10] (such as traffic volume and traffic speed) but also includes many factors. Therefore, a single parameter used to describe traffic congestion states is insufficient: when the vehicle's speed is zero, it may be blocked by too many vehicles on the road that cannot move or it may also be smooth, with no vehicles driving on the road. Therefore, a light traffic flow can match two states: heavy jam and smooth. In addition, low density may characterize traffic that is smooth and may also include more trucks and other large vehicles on the road. However, comprehensive analysis of the three variables of traffic flows (traffic flow velocity, traffic flow density and traffic volume) can reflect the real situation concerning traffic jams. Our purpose is to evaluate the traffic jam degree of different roads based on a multiple-attributes index. Thus, we introduce three variables of traffic flows to evaluate the degree of traffic jams. However, to consider addressing data of multidimensional mixed attributes and obtain data clustering levels, we use grey relational analysis theory.

Traffic flows are characterized by three properties [11]: traffic flow velocity, traffic flow density and traffic volume. Traffic flow velocity indicates the average speed of vehicles on the road, in units of km/h. The traffic flow density, namely, the density of vehicles, indicates the number of vehicles per unit length that the road contains. The traffic volume is defined as the number of vehicles traveling through a certain road section in unit time. This paper uses these three properties of traffic flows to judge the traffic congestion state.

Definition 1. *Assume the analysis domain data set* $X = \{X_i | X_i = (X_{i1}, X_{i2}, X_{i3})\}$ $i \in N$ *as the comparative object set, where* X_i *represents the ith data object, each of which has three attributes: traffic flow velocity, traffic flow density, and traffic volume.*

Definition 2. *Suppose a reference standard array set as* Y $=$ $\{Y_i|Y_i = (Y_{i1}, Y_{i2}, Y_{i3})\}$ $(i = 1, 2, \ldots, p)$ *(referenced object set). We regard the traffic flow information (when the road is in the smooth state or vehicles are traveling at the free stream velocity) as a reference standard array set.*

Each group of a reference sequence can obtain a clustering result when combined with a comparative object set by the grey relational clustering method; therefore, p groups of referenced arrays will produce p clustering results. In addition, the group of referenced standards extracted from the data is used for clustering when combined with comparative object sets, which can produce a clustering result. Thus, over the whole process, we can obtain $p + 1$ clustering results.

Definition 3. *The evaluation information M is regarded as the output information of the grey relational clustering system S, and F is defined as the evaluation function. Then, the relationship between S and M is denoted as*

$$\begin{cases} S = ((X, Y), G) \\ F : X \times Y \rightarrow M \end{cases} \tag{1}$$

where G represents the grey relational similar matrix. The evaluation function F of the above model leverages the decision table system of rough set theory to weigh the degree of contribution of the cluster members to the clustering results. F is also called the grey relational membership function, whose inputs are X and Y and output is M, which reflects the similarity between elements inside a class and the similarity between classes.

In this paper, we propose the GMRC algorithm oriented to multidimensional attribute data to judge clustering rank priority. The method's procedure includes pre-processing data, transforming the data into matrix form, setting the threshold, and filtering and deleting abnormal data objects. Next, we cluster analysis domain data objects via grey relational clustering analysis and then apply weighted decision analysis from rough set theory. Finally, the clustering results are calculated using probability theory.

4. GMRC Algorithm

The architecture of the GMRC algorithm consists of the following phases (Figure 2):

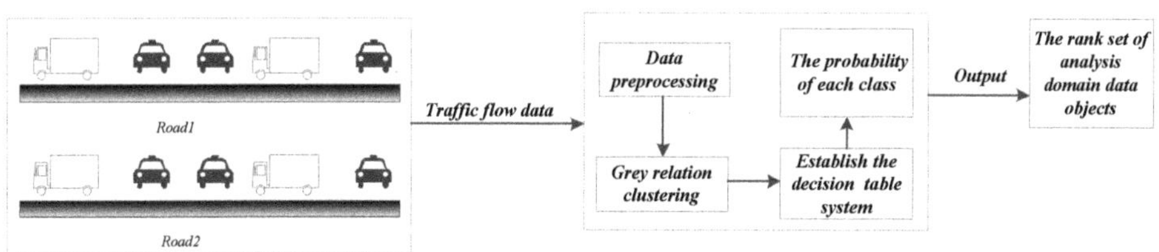

Figure 2. The architecture of the grey relational membership degree rank clustering algorithm (GMRC).

According to the index feature of the three properties of the traffic flow data, the optimal referenced standard is extracted from the data set of the analysis domain. Then, the optimal referenced standard and referenced array set are used to calculate the grey relational degree in combination with the dataset of the analysis domain. Subsequently, the grey relational matrix can be obtained, and the cluster members can be calculated. Next, we use rough set theory, where cluster members are applied, and build the decision information table to complete the fusion for the weighting of the cluster members, according to which we need to calculate the rank of each data object using probability theory. We regard the first category as the highest priority, which means that the data object is closer to the referenced standard and simultaneously that the object is better, corresponding to the lightest traffic congestion degree.

4.1. Grey Relational Clustering Steps

4.1.1. Extracting Optimal Referenced Standard according to the Characteristics of Traffic Flow Data

Because the problem of clustering analysis is solved based on a given indicator system, it is very important to choose the appropriate indicator to achieve reasonable and appropriate clustering. There are three properties for each X_i from a data source, and based on the indicator attribute of the data object, the optimal reference standard X_0 can be extracted from the data source. Specific instructions are as follows.

We know that the three properties units are different; therefore, we first need to convert the data to the same format. "↑" indicates that the property value is greater and thus better; "↓" indicates the opposition. (a, b), where a and b are numbers, indicates that, if the property value of a data object is in this interval, the value will be better.

Extract the optimal referenced standard X_0 according to the characteristics of the three properties, and its expression is $X_0 = \{X_{01}, X_{02}, X_{03}\}$, which is described as follows:

Considering the characteristic of the traffic flow velocity as "↑", we use

$$X_{0j} = \max X_{ij} \ (i \in N, j = 1) \tag{2}$$

Considering the characteristic of the traffic flow density as "↓", we use

$$X_{0j} = \min X_{ij} \ (i \in N, j = 2) \tag{3}$$

Considering the characteristic of the traffic flow volume as "↓", we use

$$X_{0j} = \left\{ X_{ij} \middle| \min \left| X_{ij} - \frac{(a+b)}{2} \right| \ (i \in N, j = 3) \right\} \tag{4}$$

4.1.2. Data Normalization Processing

Because there are different types of data, the units are also different. According to the characteristics of the properties, the data of the analysis domain are processed using different measures, and the data are compressed to $(0, 1)$. The processing is as follows:

For the traffic flow velocity of the whole data set, we use

$$X_i(j) = \left| \frac{X_{ij} - X_{j\,min}}{X_{j\,max} - X_{j\,min}} \right| \ (i = 0, 1, \ldots, n, j = 1) \tag{5}$$

For the traffic flow density of the whole data, we use

$$X_i(j) = \left| \frac{X_{ij} - X_{j\,max}}{X_{j\,max} - X_{j\,min}} \right| \ (i = 0, 1, \ldots, n, j = 2) \tag{6}$$

For the traffic volume of the whole data set, we use

$$X_i(j) = \begin{cases} 1 - \left| \frac{X_{ij} - X_{0j}}{X_{0j}} \right|, & \left| \frac{X_{ij} - X_{0j}}{X_{0j}} \right| \leqslant 1 \ and \ i = 0, 1, \ldots, n, j = 3 \\ 0, & \left| \frac{X_{ij} - X_{0j}}{X_{0j}} \right| > 1 \ and \ i = 0, 1, \ldots, n, j = 3 \end{cases} \tag{7}$$

X_{ij} is the original matrix element in the above Equations (5)-(7) and is normalized to $X_i(j)$, where $X_{j\,max}$ represents the maximum of the *jth* column and $X_{j\,min}$ represents the minimum of the *jth* column, and anything inside of braces is the limiting condition. After normalization, we obtain $(p + 1)$ matrices, namely, $A_0, \ A_1, \ldots, \ A_p$ where A_0 can be acquired by normalizing the optimal referenced standard X_0 and the comparative object set. Meanwhile, $A_1, \ A_2, \ldots, \ A_p$ can be acquired by normalizing the referenced array set and the comparative object set, where A_0 and A_p are defined as

$$A_0 = \begin{bmatrix} X_1\,(1) & X_1\,(2) & X_1\,(3) \\ X_2\,(1) & X_2\,(2) & X_2\,(3) \\ \cdots & \cdots & \cdots \\ X_n\,(1) & X_n\,(2) & X_n\,(3) \\ X_0\,(1) & X_0\,(2) & X_0\,(3) \end{bmatrix} \qquad A_p = \begin{bmatrix} X_1\,(1) & X_1\,(2) & X_1\,(3) \\ X_2\,(1) & X_2\,(2) & X_2\,(3) \\ \cdots & \cdots & \cdots \\ X_n\,(1) & X_n\,(2) & X_n\,(3) \\ Y_p\,(1) & Y_p\,(2) & Y_p\,(3) \end{bmatrix} \tag{8}$$

The last row of the matrix A_0 is normalized to the optimal referenced standard sequence, and the last row of the matrix A_p is the *pth* normalized referenced standard array.

4.1.3. Calculating Grey Relational Degree and Generating Grey Relational Similar Matrix

(1) The grey relational degree reflects the high degree between two comparative objects. For example, we focus on calculating the grey relational degree of the matrix A_0. The formulas are as follows (Equations (9) and (10)):

$$\gamma_{0i(k)} = \frac{\min\limits_i \min\limits_k |X_0\,(k) - X_i\,(k)| + \sigma \max\limits_i \max\limits_k |X_0\,(k) - X_i\,(k)|}{|X_0\,(k) - X_i\,(k)| + \sigma \max\limits_i \max\limits_k |X_0\,(k) - X_i\,(k)|} \qquad (k = 1,2,3) \tag{9}$$

$$\gamma_{0i} = \frac{1}{m} \sum_{k=1}^{m} \gamma_{0i}\,(k) \tag{10}$$

where σ is the resolution coefficient, which has a range of 0 to 1. Generally, we assume that σ is 0.5. $\gamma_{0i}\,(k)$ is the correlation coefficient between X_i and X_0 at the *kth* point. The grey relational degree is expressed by $\gamma_{0i}\,(k)$, where X_0 is regarded as the referenced sequence and X_i is regarded as the comparative sequence. Similarly, we can obtain γ_{ij} when X_i is regarded as the referenced sequence and X_j is regarded as the comparative sequence. Then, X_1, $X_2,\ldots$, X_n, X_0 are regarded as the referenced sequence; meanwhile, the $(n + 1)$ sequences are regarded as comparative sequences (the $(n + 1)$ sequences are not only referenced sequences but also comparative sequences). Finally, we calculate the grey relational degree matrix $\Gamma^0_{(n + 1) \times (n + 1)}$ according to the grey relational analysis, where $\Gamma^0_{(n + 1) \times (n + 1)}$ is obtained based on A_0 and consists of any γ_{ij} $(i = 1, 2,\ldots,n,n + 1, j = 1,2,3)$. Similarly, we can calculate the grey relational degree matrices Γ^1, $\Gamma^2,\ldots,\Gamma^p$ based on $A_0, A_1,\ldots, A_p$.

(2) Calculate grey relational similar matrix G

We calculate the similarity elements $g_{ij} = \left(\gamma_{ij} + \gamma_{ji}\right)/2$ in G. G_0, for example, is the grey relational similar matrix obtained when X_0 is regarded as the referenced sequence, and the $(n + 1)th$ row grey relational similarity elements of G_0 are calculated when X_0 is regarded as the optimal referenced sequence. Similarly, we can obtain $p + 1$ grey relational similar matrices: $G_0, G_1,\ldots, G_p$.

4.1.4. Grey Relational Degree Clustering Analysis

Based on G, *i.e.*, the grey relational similar matrix, we construct the maximal relational tree, which consists of the values of the last row elements ordered in G. This is shown in Figure 3.

Figure 3. Maximal relational tree.

Based on G_0, we can obtain g_{ij}, which is also called the closeness degree between X_i and X_j. The greater g_{ij} is, the closer X_i and X_j are; in contrast, X_i and X_j are further away for decreased

g_{ij}. Based on Figure 3, the maximal relational tree with closeness degrees is generated, as shown in Figure 4, where a_i represents the closeness degree between X_i and X_j.

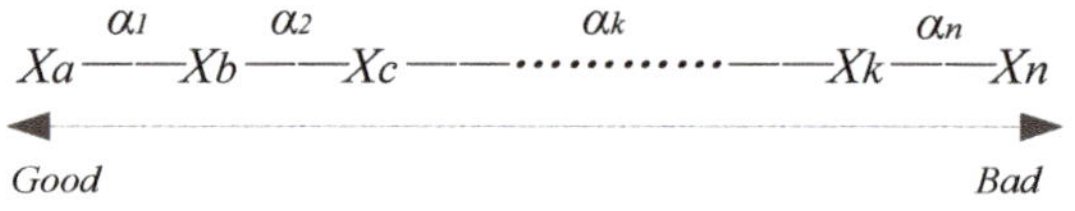

$$Xa \overset{\alpha_1}{———} Xb \overset{\alpha_2}{———} Xc ——— \cdots\cdots\cdots\cdots ——— Xk \overset{\alpha_n}{———} Xn$$

$$\blacktriangleleft \text{————————————————————————} \blacktriangleright$$

Good   Bad

Figure 4. Maximal relational tree with closeness degrees.

Based on Figure 4, we set the isolated point coefficient λ, which is in the interval $(0, 1)$. We cut the tree when the closeness degree is less than λ and when adjacent branches exhibit substantial differences. Therefore, the disconnected tree is used to make its connected branches form k levels of clusters along the horizontal. We consider the closest branch as the first level and the loosest branch as the kth level (If $k \geqslant 4$, we also classify the fifth branch, the sixth branch, the seventh branch, and even the kth branch as the fourth rank. This is to say that smooth corresponds to the first rank and that heavy jam corresponds to the fourth rank). Thus, in this way, we can obtain a cluster member of G. Therefore, we use Γ^0, $\Gamma^1, \ldots, \Gamma^p$ to compute $p + 1$ grey relational similarity matrices G_0, G_1, $\ldots$, G_p, from which we can find the closeness relationship among each object. Therefore, we can obtain $p + 1$ clustering results, namely, cluster members.

4.2. Establishing Evaluation Function for Grey Relational Clustering System

According to the above initial clustering results, we use rough set theory to establish the decision table system that is applied to weight the contribution of cluster members to the clustering results and give weights to the cluster members.

4.2.1. Describing How to Establish the Decision Table System

First, the optimal referenced standard and the referenced standard set are combined with the comparative object set through the grey relational clustering method to construct the decision table system $F = \langle U, C, D, V, f \rangle$,where $U = \{X_1, X_2, \ldots, X_n\}$ represents analysis domain data; $C = \{c_1, c_2, \ldots, c_p\}$ are conditional attributes and are cluster members formed by the referenced standard set; $D = \{d\}$, as the decisional attribute, is the cluster member obtained using the optimal referenced standard; $V = V_C \cup V_D$, $V_C = \cup V_{c_h}$, $c \in C$ are the range of the set of traffic flow properties, where V_{c_h} represents the level in cluster member c_h $(h = 1, 2, \ldots, p)$; f represents the evaluation function, $f : U \times (C \cup D) \rightarrow V$; and $f(X_i, c_h) \in V_{c_h}$ represents the level of X_i in cluster member c_h.

4.2.2. Calculating Information Entropy

In the decision table system, the information entropy weight $I(c_h, D)$ indicates how important the cluster member c_h (conditional information) is for result D (decisional information) when the optimal referenced standard is chosen to calculate the cluster member. According to the information entropy of rough set theory, $I(c_h, D)$ is described as follows:

$$I(c, D) = H(D) - H(D|\{c\}) \tag{11}$$

$$H(c) = -\sum_{i=1}^{k} P(RC_i) \, log \, (P(RC_i)) \tag{12}$$

$$H(D|\{c\}) = -\sum_{i=1}^{k} P(RD_i|RC_i) \, log \, (P(RD_i|RC_i)) \tag{13}$$

$$P(RC_i) = \frac{|RC_i|}{|X|} \tag{14}$$

$$P\left(RD_i|RC_i\right) = \frac{|RD_i \cap RC_i|}{|RC_i|} \tag{15}$$

where $i = 1, 2, \ldots, k$ (k is the number of clusters), RC_i represents the *ith* divided cluster of cluster member c, $|RC_i|$ represents the number of elements in the *ith* cluster, and $|X| = n$. A larger conditional attribute c is more important to the decisional information D. In addition, $H(c)$ and $H(D|\{c\})$ are determined by the conditional information entropy of rough set theory. Thus, the relative weight of each cluster member can be determined, and a more important cluster member corresponds to a greater weight.

4.3. Calculating the Level of Clustering Membership of Data Objects

Step 1: Calculate the importance of the attribute information entropy $E_h = I\left(c_h, D\right)$ for each cluster member c_h in the decision system, where $h = 1, 2, \ldots, p$.

Step 2: Set the relative weight of each cluster member:

$$\omega_h = E_h / \sum_{h=1}^{p} E_h \tag{16}$$

Step 3: Use probability theory to calculate the probability of each data object emerging in every clustering based on the relative weights to choose the level whereby the probability is maximized. Furthermore, obtain the final clustering results. In addition, data object X_i belonging to the *jth* level ($j = 1, 2, \ldots, k$) is defined as

$$P\left(X_i^j\right) = \sum_{\substack{h=1 \\ M_{ik}=j}}^{p} \omega_h \tag{17}$$

where M_{ik} represents the level of data object X_i in cluster member c_h, which has been computed in grey relational clustering. Thus, the grey relational membership degree level of X_i can be expressed as:

$$Level\left(X_i\right) = \{j|\max_{j=1 \rightarrow k}\left(P(X_i^j)\right)\} \tag{18}$$

The final result is $C = \left\{C^1, C^2, \ldots, C^k\right\}$, where C^k includes all data objects whose grey relational membership degree level is the *kth* level:

$$C^k = \{X_i|Level\left(X_i\right) = k, X_i \in X\} \tag{19}$$

4.4. GMRC Algorithm Detail Description

In this paper, we study the problem of multidimensional-attribute information clustering for traffic flow and propose the GMRC algorithm. First, we transform the dataset into matrix form, extract the optimal referenced standard from the dataset, and then perform the normalized processing to eliminate the effects of different units. Furthermore, we obtain the preliminary clustering results according to grey relational theory analysis. Finally, we build a decision table system to calculate the relative weight for each cluster member. The algorithm is described as follows.

Input: Analysis domain data set $X = \{X_i|X_i = (X_{i1}, X_{i2}, X_{i3})\}$ $i \in N$,

           Referenced array set $Y = \{Y_i|Y_i = (Y_{i1}, Y_{i2}, Y_{i3})\}$ ($i = 1, 2, \ldots, p$).

Output: The level (rank) set of analysis domain data objects

Algorithm 1: GMRC (X,Y)

1	Level = null; Weight = null; Member = null; Entropy = null;
	//Initialize sets Level, Weight; Matrix Member, Entropy
2	Data_ Preprocessing (X,Y); // Data pre-processing
3	X_0 = ExtractOptimal (X); // Extract the optimal referenced standard
4	S = Normalization (X,Y); // Normalization processing
5	T = MaxRelTrees (S) // Construct the maximum relational trees
6	T' = ClosenessTrees (T) // Construct the maximum relational tree with closeness degree
7	$Member_0$ = GreyCluster (X,X_0); // Regard X_0 as referenced standard to compute cluster member
8	Foreach (Y_i in Y)
9	$Member_i$ = GreyCluster (X,Y_i); // Regard Y_i as referenced standard to compute cluster member
10	End
11	F = DecisionSystem (Members); // Establish the decision table system F
12	Entropy = CalculateEntropy (F); // Calculate the information entropy for each cluster member
13	Weight = CalculateWeight (Entropy); // Calculate relative weights for each cluster member
14	CalculateLevel (X); // Calculate membership degree level of X_i in X

The above steps of the algorithm are described as follows:

Step 1: Initialize the parameters, pre-process traffic flow data, set the threshold to filter and delete abnormal data objects (lines 1 and 2).

Step 2: According to the features of the three properties of the traffic flow data, extract the optimal referenced standard from the analysis domain data (line 3).

Step 3: Normalize the analysis domain data set in combination with the referenced standard sequences (line 4).

Step 4: Compute the grey relational degree of the corresponding matrix; further, determine the grey relational similarity matrices and then construct the maximum relational trees based on the $(n + 1)th$ row elements of those matrices (line 5).

Step 5: Based on Step 4, construct the maximum relational tree with closeness degrees (line 6).

Step 6: Compare the closeness degree between data objects, cut off the tree when the closeness degree is less than λ and adjacent branches exhibit large differences, and then obtain k levels of clustering results. Similarly, we obtain in total $p + 1$ cluster members from the $p + 1$ referenced arrays (lines 7–10).

Step 7: Establish the decision table system based on p cluster members as conditional attributes obtained from the referenced standard array set. In addition, the only cluster member obtained from the optimal referenced standard (line 11) is regarded as the decisional attribute.

Step 8: Compute the information entropy of cluster members for decision making, which is used to weigh the contribution of each cluster member to the clustering results (line 12).

Step 9: Calculate the weight of each cluster member (line 13).

Step 10: Calculate the probability of each data object emerging in every clustering; then, choose the level when the probability is maximized. Furthermore, obtain the final clustering results (line 14).

4.5. Grey Relational Membership Function

The performance of the traditional clustering results depends on the distance between elements inside classes and the distances between classes. Shorter distances between elements inside a class indicate better classes; conversely, longer distances between classes indicate better classes [12]. Our purpose for clustering is to obtain the membership degree rank of classes and to judge the rank of data objects. Thus, we construct a membership function reflecting the similarity between elements inside a class and the similarity between classes based on the grey relational similarity degree. Assume that $\gamma_{X,Y}$ is denoted as the grey relational similarity degree between object X and object

Y, and $C = \left\{ C^1, C^2, \ldots, C^k \right\}$ represents divided clusters. S_C, which is used to weigh the division C, is defined as

$$S_c = \frac{1}{k} \sum_{i=1}^{k} \frac{1}{|C^i| \times |C^i|} \sum_{X,Y \in C^i} \gamma_{XY} + \frac{1}{k(k-1)} \sum_{i=1}^{k} \left(\sum_{i=1, l \neq i}^{k} \frac{1}{|C^i| \times |C^l|} \sum_{X \in C_i, Y \in C^l} \gamma_{XY} \right) \qquad (20)$$

5. Experimental Results and Analysis

In this paper, we collected traffic data for Nanjing trunk roads, which connect the main commercial areas and which include high traffic volumes and high-density residential areas. These roads need to be cleared in a timely manner before traffic jams can occur. Generally speaking, people's daily routines are very regular [13]: go to work in the morning and go home at dusk. However, this can quickly lead to traffic congestion during rush hours. A previous study noted that a single parameter, such as velocity, used to describe traffic congestion states is insufficient. To obtain detailed knowledge for judging traffic congestion, comprehensive analysis of the three variables of traffic flows for determining the traffic flow state can be used to reflect the real conditions of the road for predicting traffic jams [14]. To test our algorithm, we experimentally collected traffic flow data from 50 monitoring points along Nanjing's trunk roads during the time periods of approximately 7:00–9:30 a.m. and 4:30–7:00 p.m., which correspond to the rush hours. In addition, we chose 30 drivers with more than five years of driving experience as testers and watched their vehicle driving videos to obtain their evaluation of the traffic flows' four states (smooth, light jam, jam, and heavy jam) [15]. Then, we evaluated the clustering results to validate the accuracy of our algorithm compared with other clustering methods such as the K-means algorithm and the FCM algorithm.

In this experiment, we assumed that the resolution coefficient σ is 0.5 and that the number of levels is 4. Because the algorithm is stochastic in nature, the average results of 20 tests for each algorithm on each dataset are used as the experimental results. Table 1 shows the corresponding information entropy of the four cluster members obtained by the primary clustering data samples, namely, the relative weights of the cluster members. To verify the performance of our algorithm, six sample points were randomly selected from the dataset, as shown in Table 2. In Table 2, we find that the clustering of the traffic state is most closely related to traffic flow velocity; however, the velocity cannot fully determine the traffic state. For example, the traffic flow speed in the 459th group is 40.3 km/h, and the state is a light jam. However, the traffic flow speed in the 812th group is 45.5 km/h, which is greater than 40.3 km/h, but the state is a jammed state. This phenomenon occurs mainly because the traffic volume of the latter is greater than that of the former. The three properties of the traffic flow should be comprehensively considered rather than only the velocity.

Table 1. Relative weights of cluster members.

Cluster Members	Relative Weights
C1	0.3746
C2	0.2644
C3	0.2128
C4	0.2481

Table 2. Random samples.

Samples	Traffic Flow Velocity	Traffic Flow Density	Traffic Volume	Results
15	66.0	58.2	16.2	Smooth
35	35.2	70.3	25.6	Jam
106	50.8	66.9	23.9	Smooth
459	40.3	68.3	31.2	Light Jam
689	10.8	81.9	41.7	Heavy Jam
812	45.5	63.6	58.9	Jam

5.1. Complexity Analysis

We suppose k levels, n data objects, m dimensions, and p referenced standard sequences. The complexity of the GMRC algorithm is described in the following.

5.1.1. Time Complexity Analysis

The time for constructing the initial matrix is $O(m \times n)$. Extracting the optimal referenced standard requires access to all elements in the initial matrix, and the time complexity is also $O(m \times n)$. The time complexity of the matrix normalization is $O(p \times n \times m)$. The grey relational degree calculation requires access to all the elements in n grey relational similarity degree matrices; therefore, its time complexity is approximately $O(n \times m \times p \times n)$. The time complexity of sorting the $(n+1)th$ row grey relational degree elements for p grey relational similarity degree matrices is $O(p \times n^2)$. Calculating the information entropy for the cluster members needs $O(p \times k)$. The time needed to calculate the clustering membership degree analysis domain data is $O(n \times p \times k)$. According to the above analysis, the average time complexity of the GMRC algorithm is $O(k \times m \times p \times n^2)$.

5.1.2. Space Complexity Analysis

The complexity of analyzing the domain initial matrix is $O(m \times n)$ in space, and the complexity of analyzing the similarity degree matrix with space complexity is $O(p \times n^2)$. In addition, the space complexity of determining the grey relational membership degree in the GMRC algorithm is $O(k \times p)$. Therefore, the total space complexity of the algorithm is $O(k \times p \times n^2)$.

5.2. Impact of Isolated Point Coefficient λ on the Clustering Results

From Figure 5, we can find that, based on different numbers of data samples, the grey relational membership function value gradually increases with increasing λ. This is because with increasing λ, the grey relational similarity degree between the elements inside classes increases, and this accounts for the dominant position. Clearly, when λ is between 0.84 and 0.87, the function value is maximized. Then, as λ continues to gradually increase, the function value decreases. This is because the grey relational similarity degree between classes occupies the dominant position. From the perspective of the grey relational similarity degree between data objects, we analyze the closeness degree inside classes and among classes and fully utilize the multi-dimensional information feature and the overall change in the three properties to better describe the closeness degree between data objects. Therefore, in the next experiment, the range of λ is set as (0.84, 0.87).

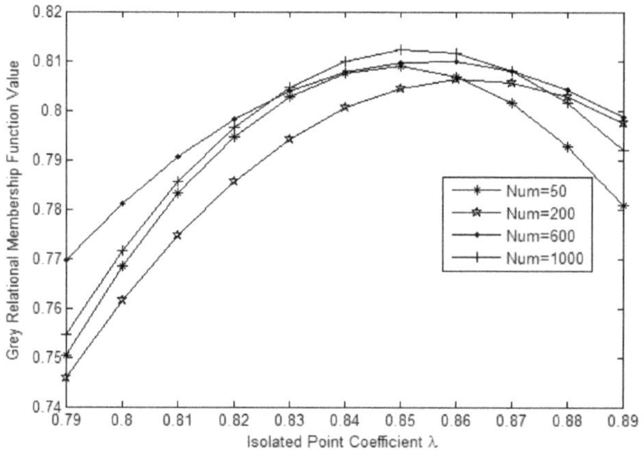

Figure 5. Grey relational membership function as a function of λ.

5.3. Comparison with Other Algorithms

Figure 6 illustrates the accuracy rate of the GMRC, K-means and FCM clustering algorithms in each class. We find that the accuracy of GMRC in each class and under different λ is higher than that of the K-means and Fuzzy algorithms. This is because the GMRC algorithm uses the data attribute index feature to compute the grey relational similarity degree and takes the quality of cluster members into consideration. Thus, the algorithm effectively improves internal class similarity to achieve ranked clustering.

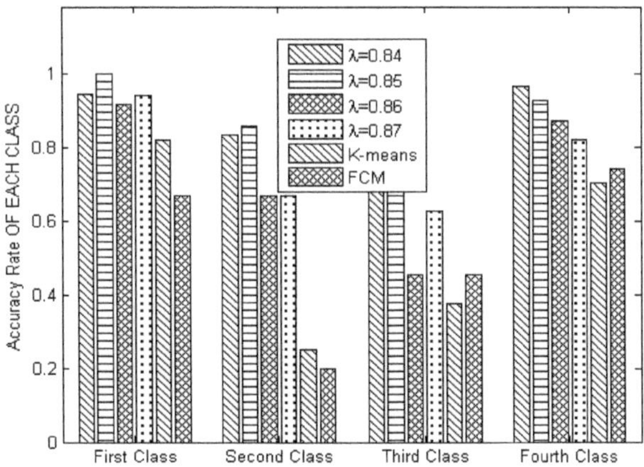

Figure 6. Comparison of accuracy rate of each class among GMRC, K-means and FCM algorithms.

Figure 7 shows the average accuracy of our GMRC algorithm under different λ and that of the K-means and FCM algorithms. From Figures 6 and 7 we can conclude that the average accuracy of the GMRC algorithm is higher than that of the K-means and FCM algorithms. In addition, the average accuracy of the GMRC algorithm is 24.9% higher than that of the K-means algorithm and 30.8% higher than that of the FCM algorithm. In addition, our new algorithm exhibits better stability. Because our algorithm does not need to randomly choose the initial center point, as in the K-means algorithm, the stability of the algorithm is not affected by the stochastic nature of the algorithm.

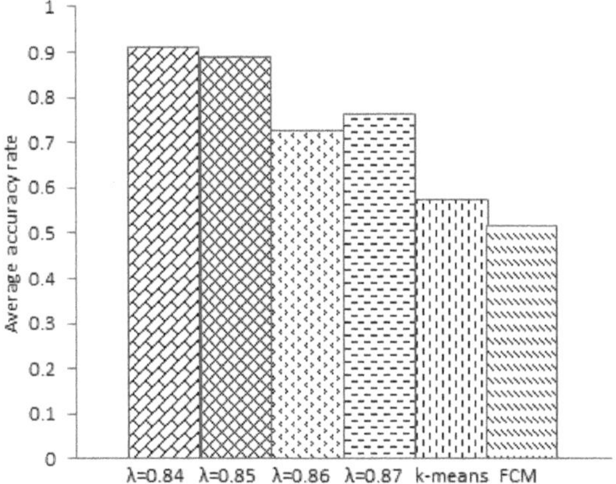

Figure 7. Comparison of average accuracy rate among the GMRC algorithm with different λ and the K-means and FCM algorithms.

6. Conclusions

Judging traffic congestion states is the premise and basis for dynamic traffic congestion warning, traffic guidance, actively avoiding traffic congestion and ensuring smooth roads. However, traffic jams are usually judged by experience. This paper collects traffic flow data and provides a more effective judgment method. We introduce both grey relational analysis and rough set theory to the GMRC algorithm and weigh the membership degree of data object clustering using comprehensive information about the data. In this process, we construct the maximum relational tree with closeness degree and compare the closeness degree between data objects. We cut off the tree when the closeness degree is less than λ and when adjacent branches exhibit large differences. Consequently, we obtain $p + 1$ cluster members. Next, we establish a decision table system based on p cluster members as conditional attributes obtained from referenced standard array sets. Then, we calculate the probability of each data object emerging in every clustering, choose the rank when the probability is maximized, and finally obtain the final clustering results. Thus, our algorithm fills the gaps present in the literature whereby the K-means and FCM algorithms cannot differentiate which rank a clustering belongs to. The experimental results show that the proposed algorithm, which takes the characteristics of the multidimensional data object attributes into consideration, is a superior algorithm. Next, we plan on applying grey relational similarity to other algorithms and to consider reducing the computational complexity of the algorithm.

Acknowledgments: This research was performed in cooperation with the Institution. The research is support by the National Natural Science Foundation of China (No. 61572260, No. 61373017, and No. 61572261), the Peak of Six Major Talent in Jiangsu Province (No. 2010DZXX026), the China Postdoctoral Science Foundation (No. 2014M560440), the Jiangsu Planned Projects for Postdoctoral Research Funds (No. 1302055C), the Scientific & Technological Support Project of Jiangsu Province (No. BE2015702), the Jiangsu Provincial Research Scheme of Natural Science for Higher Education Institutions (No. 12KJB520009), and the Science & Technology Innovation Fund for Higher Education Institutions of Jiangsu Province (No. CXZZ11-0405). The authors are grateful to the anonymous referee for a careful review of the details and for their helpful comments, which improved this paper.

Author Contributions: All the authors conceived of and designed the study. Furthermore, Yingya Zhang designed the GMRC algorithms presented in this paper and produced the results. Ning Ye provided guidance on modeling, Ruchuan Wang provided guidance on theory and publication fee, and Malekian provided guidance on simulation.

Conflicts of Interest: The authors declare no conflict of interest.

References

1. Courtney, R.L. A broad view of its standards in the U.S. In Proceedings of the IEEE Conference on Intelligent Transportation Systems, Boston, MA, USA, 9–12 November 1997; pp. 529–536.
2. Arnold, E.D., Jr. Congestion on Virginia's Urban Highways. Available online: http://ntl.bts.gov/DOCS/arnold.html (accessed on 17 April 1998).
3. Mok, P.Y.; Huang, H.Q.; Kwok, Y.L.; Au, J.S. A robust adaptive clustering analysis method for automatic identification of clusters. *Pattern Recognit.* **2012**, *45*, 3017–3033. [CrossRef]
4. Yu, B.; Zhou, Z.; Xie, M. New grey comprehensive correlation degree model and its application. *Technol. Econ.* **2013**, *32*, 108–114.
5. Hu, F.; Xia, G.-S.; Wang, Z.; Huang, X.; Zhang, L.; Sun, H. Unsupervised feature learning via spectral clustering of multidimensional patches for remotely sensed scene classification. *IEEE J. Sel. Top. Appl. Earth Obs. Remote Sens.* **2015**, *8*, 2015–2030.
6. Li, K.; Li, P. Fuzzy Clustering with generalized entropy based on neural network. In *Unifying Electrical Engineering and Electronics Engineering*; Springer: New York, NY, USA, 2014; pp. 2085–2091.
7. Zhang H, R.; Zhang, F. The traditional K-means clustering algorithm research and improvement. *J. Xianyang Norm. Univ.* **2010**, *25*, 138–144.
8. Zheng, Y.; Jeon, B.; Xu, D.; Wu, Q.M.J.; Zhang, H. Image segmentation by generalized hierarchical fuzzy C-means algorithm. *J. Intell. Fuzzy Syst.* **2015**, *28*, 961–973.
9. Li, K.; Cui, L. A kernel fuzzy clustering algorithm with generalized entropy based on weighted sample. *Int. J. Adv. Comput. Res.* **2014**, *4*, 596–600.

10. Wen, H.; Sun, J.; Zhang, X. Study on traffic congestion patterns of large city in China taking Beijing as an example. *Procedia-Soc. Behav. Sci.* **2014**, *138*, 482–491. [CrossRef]

11. Stefanello, F.; Buriol, L.S.; Hirsch, M.J.; Pardalos, P.M.; Querido, T.; Resende, M.G.C.; Ritt, M. On the minimization of traffic congestion in road networks with tolls. *Ann. Oper. Res.* **2015**. [CrossRef]

12. Guan, X.; Sun X, W.; He, Y. A novel feature association algorithm based on grey correlation grade and distance measure. *Radar Sci. Technol.* **2013**, *4*, 363–367, 374.

13. He, H.; Tang, Q.; Liu, Z. Adaptive correction forecasting approach for urban traffic flow based on fuzzy C-Mean clustering and advanced neural network. *J. Appl. Math.* **2013**, *2013*, 633–654.

14. Lu, X.; Song, Z.; Xu, Z.; Sun, W. Urban traffic congestion detection based on clustering analysis of real-time traffic data. *J. Geo-Inf. Sci.* **2012**, *14*, 775–780. [CrossRef]

15. Elefteriadou, L.; Srinivasan, S.; Steiner, R.L.; Tice, P.C.; Lim, K. Expanded transportation performance measures to supplement level of service (LOS) for growth management and transportation impact analysis. *Congest. Manag. Syst.* **2012**, *19*, 977–991.

International Journal of
Geo-Information

Article

Smartphone-Based Pedestrian's Avoidance Behavior Recognition towards Opportunistic Road Anomaly Detection [†]

Tsuyoshi Ishikawa [‡] and Kaori Fujinami [*,‡]

Department of Computer and Information Sciences, Tokyo University of Agriculture and Technology, Naka-cho Koganei, 2-24-16 Tokyo, Japan; 14kawa244@gmail.com

* Correspondence: fujinami@cc.tuat.ac.jp; Tel.: +81-42-388-7499
† This paper is an extended version of our paper published in Pedestrian's avoidance behavior recognition for road anomaly detection in the city in the Proceedings of the the ACM International Joint Conference on Pervasive and Ubiquitous Computing and ACM International Symposium on Wearable Computers, Osaka, Japan, 7–11 September 2015.
‡ These authors contributed equally to this work.

Academic Editors: Chi-Hua Chen, Kuen-Rong Lo and Wolfgang Kainz
Received: 24 May 2016; Accepted: 28 September 2016; Published: 3 October 2016

Abstract: Road anomalies, such as cracks, pits and puddles, have generally been identified by citizen reports made by e-mail or telephone; however, it is difficult for administrative entities to locate the anomaly for repair. An advanced smartphone-based solution that sends text and/or image reports with location information is not a long-lasting solution, because it depends on people's active reporting. In this article, we show an opportunistic sensing-based system that uses a smartphone for road anomaly detection without any active user involvement. To detect road anomalies, we focus on pedestrians' avoidance behaviors, which are characterized by changing azimuth patterns. Three typical avoidance behaviors are defined, and random forest is chosen as the classifier. Twenty-nine features are defined, in which features calculated by splitting a segment into the first half and the second half and considering the monotonicity of change were proven to be effective in recognition. Experiments were carried out under an ideal and controlled environment. Ten-fold cross-validation shows an average classification performance with an F-measure of 0.89 for six activities. The proposed recognition method was proven to be robust against the size of obstacles, and the dependency on the storing position of a smartphone can be handled by an appropriate classifier per storing position. Furthermore, an analysis implies that the classification of data from an "unknown" person can be improved by taking into account the compatibility of a classifier.

Keywords: road anomaly; avoidance; behavior recognition; smartphone; opportunistic sensing

---

1. Introduction

Road anomalies, such as cracks, pits, puddles and fallen trees, are generally identified from citizen reports and are repaired by administrative entities. In most cases, the reports are made by telephone or e-mail, which makes it difficult for the administrative entities to identify the location of the anomaly. To address this issue, administrative entities and third parties are attempting to provide smartphone-based applications that accept text and/or image reports with location information [? ?]. Such human-centric sensing is often called participatory sensing [?]. Although the success of these applications depends on people actively reporting tasks, very few of the citizens who downloaded these applications have actually reported anomalies [?]. Therefore, we propose a method to detect road anomalies implicitly based on opportunistic sensing. Opportunistic sensing is another human-centric

sensing paradigm, in which the data collection process is automated without any user involvement [?
].

To detect road anomalies, we focus on avoidance behaviors. Recognizing avoidance behaviors
and aggregating events with locations can help to generate automatic anomaly reports. Automatic road
anomaly detection techniques for cars and bikes have already been proposed [? ? ? ? ?], but these
cases deal with relatively large movements. In contrast, we consider that pedestrians' avoidance
behaviors are too slight to make the adaptation of existing methods acceptable. The contributions of
this article are as follows:

- A smartphone-based road anomaly detection system is presented, in which obstacle avoidance
  behaviors are categorized into three classes. The three classes include: (1) returning to the
  same line in the vicinity of avoiding an obstacle; (2) going straight after avoiding an obstacle;
  and (3) reversing his/her course; which may indicate the impact of the obstacle on pedestrians.
  The three classes may indicate the severity of obstacles, which would be helpful for an
  administrative entity to plan a repair schedule.
- Twenty nine classification features are defined based on the characteristics of the azimuth change
  of each class. The relevance of the features is evaluated.
- We extensively analyze the effects of various factors on the recognition performance.
  This includes the individuals who provide data for training classifiers and the position of sensors
  (i.e., smartphones) on their bodies, as well as the size of target obstacles.

An initial decision on the position and the class of an obstacle is made on the smartphone side
against a stream of sensor data, while the collected information from a number of pedestrians is
utilized to make the final decision. Low-power operation and server-side processing are beyond the
scope of this article. Furthermore, we do not deal with a method of distinguishing a normal behavior,
e.g., walking along a curved road, from avoidance behavior. Instead, we focus on classifying a data
segment of avoidance behavior into one of six (three classes × right and left turns) classes.

The remainder of this article is organized as follows. In Section ??, related work is presented.
Section ?? shows the system overview, followed by offline experiments in Section ??. Finally, Section ??
concludes the article. Note that, in [?], we proposed the basic idea of smartphone-based road
anomaly detection. This article has extensions in the following points: a section of related work is
added in order to clarify the uniqueness of the work (Section ??); the overall system ideas are presented,
including not only the local processing on the smartphone, but also the server side processing to filter
out erroneous detection from the smartphone side (Section ??); the detail of avoidance behavior
recognition (on the smartphone) is described, including how the raw azimuth data stream is processed
into the final avoidance event and detail definition of features (Section ??); experiments were carried
out with different conditions, i.e., a type of behavior "straight" was excluded, because we considered it
could be done in the preprocessing stage; and extensive analyses about person dependency (Section ??),
sensor-storing position dependency (Section ??) and robustness to unknown obstacle size (Section ??)
were undertaken.

2. Related Work

Motorcycles and other vehicles are often used as a method of automatic road anomaly detection
and unsafe behavior detection [? ? ? ? ? ? ? ?]. Thepvilojanapong et al. [?] and Kamimura et
al. [?] proposed a method using a smartphone-mounted accelerometer and gyroscope to detect
driving activities, such as turning left or right, going forward or bicycles and motorcycles going
past nearby cars. Iwasaki et al. [?] proposed a method to recognize road characteristics, such
as an intersection with poor visibility and a congested road based on the bicycle riding behavior.
Additionally, a special sensor unit that measures rudder angle and velocity was developed for bicycle
riders to detect hazardous locations [?]. The vertical displacements of vehicles passing over bumps
and potholes are often subject to monitoring in the case of cars [? ? ? ? ?]. In contrast, Chen et

al. recognized driving factors that cause horizontal displacements, such as a lane change, S-shaped curved road, turning or L-shaped curved road [?]. In the above-mentioned work, all but [?] utilized a smartphone-mounted accelerometer, gyroscope and/or magnetometer. This shows the possibilities of smartphones as easy-to-deploy sensors in combination with a positioning technique, e.g., GPS. Furthermore, the aggregation of data followed by proper analysis can create new types of information content for comfortable and safe transport systems. Our work shares its motivations with the above studies. However, we focus on road anomalies from the pedestrian's point of view. In addition, although the above-mentioned work on horizontal displacement [? ? ? ?] might find similar trajectories of moving objects, bicycles and cars have larger and faster movements than pedestrians, and so, applying the existing methods to our domain would be difficult.

For pedestrian-based road condition monitoring, Jain et al. proposed a shoe-based ground gradient sensing technique [?]. A sensing unit composed of a magnetometer, accelerometer and gyroscope is mounted on shoes, which collect data that detect the transitions between sidewalks and streets through the recognition of dedicated slopes. The primary motivation of those studies was to use the data to alert texting pedestrians who are about to step into the street. The slope sensing technique can also detect the vertical condition of sidewalks, e.g., bumps. Additionally, the shoes can detect turns and moving direction. These capabilities suggest that augmented shoes can be combined with our system as a sensor to detect horizontal behavior changes.

3. Avoidance Behavior Recognition

3.1. System Overview

Figure ?? illustrates the concept of the proposed system. The proposed system is designed to identify a road anomaly in an automatic manner, which consists of an avoidance behavior recognition function with location measurement on the smartphone side and aggregation and filtering functions on the server side. An avoidance behavior is recognized by measuring azimuth changes of the walking direction by a smartphone-mounted accelerometer and magnetometer. In the Android API, these two sensors are internally utilized to obtain azimuth data. The position where the avoidance event occurs is measured by a positioning technique, such as GPS. The information is sent to a database on the server side. In Figure ??, av_1, av_2 and av_3 indicate the candidates of avoidance events. Note that power consumption is a central issue for the success of opportunistic sensing from the user's point of view. To minimize the communication with a server, the processing for the avoidance event detection is carried out on the terminal side, and only events of avoidance behavior are sent to the server. Additionally, we assume the GPS receiver is activated only when an avoidance event is detected (the positional gap between the time of event detection and that of GPS-ready is also considered).

The aggregated information may contain erroneous events that are falsely recognized as avoidance behaviors, such as one person passing another person and looking behind with the smartphone terminal in his/her hand, as well as the effect of positioning error. Therefore, spatio-temporal filtering should be applied to extract only "static obstacles" on the road (e.g., [?]). In Figure ??, av_1 is such a false detection, and the system finally identifies a road anomaly near the position of av_2 and av_3. Calculating the center of the positions of avoidance behavior events is a simple solution. Additionally, a geographical information system (GIS) for map-matching the position of an event on a road can be applied. A GIS can also be utilized to eliminate an event falsely classified as avoidance, which is actually normal behavior, by reflecting the semantics of the road, i.e., identifying that a curve exists at position (x, y). This article focuses on the avoidance recognition functionality on the smartphone side. Low-power positioning and server-side processing are beyond the focus of this article.

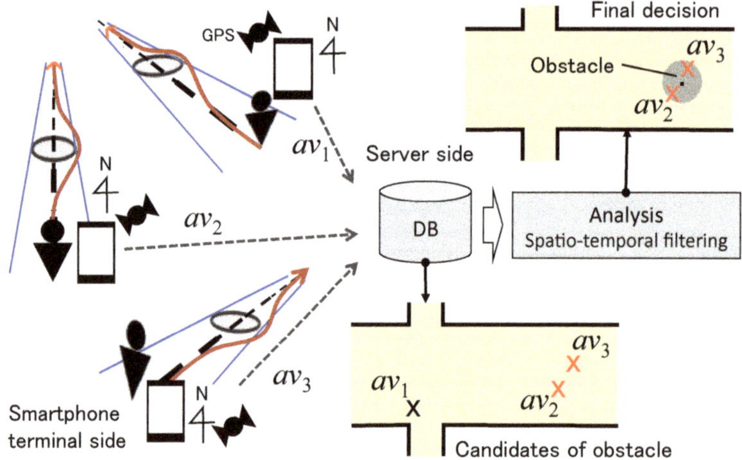

Figure 1. Concept of the automatic road anomaly reporting system.

3.2. Avoidance Behavior Modeling

We focus on detecting anomalies on the road surface, such as pits, cracks, puddles, fallen trees and landslides. These anomalies make pedestrians change their walking paths as a natural defensive behavior [?]. The avoidance behavior during walking is modeled by the combination of three elements: (1) avoiding direction (left or right); (2) going through an obstacle (avoid) or going back (return); and (3) a direction change after avoiding an obstacle. In total, six types of avoidance behaviors are defined, as shown in Figure **??**: $avoid_{LR}$, $avoid_L$, $avoid_{RL}$, $avoid_R$, $return_L$ and "$return_R$". Here, the postfix "LR" indicates, for example, that the pedestrian changes direction to the left followed by a change to the right, whereas the postfix "L" alone does not have the second change after the first change to the left. The horizontal dotted line in Figure **??** indicates the pedestrian's straight walking path. Furthermore, d represents the size of an avoidance behavior, which is primarily determined by the physical size of an obstacle, i.e., the avoidance behavior size equals the obstacle size. The perceived size may also affect the behavior or, in other words, the severity of the anomaly. We collectively call d "obstacle size" or "size of obstacle". The typical waveforms of raw azimuth signals are shown in Figure **??**.

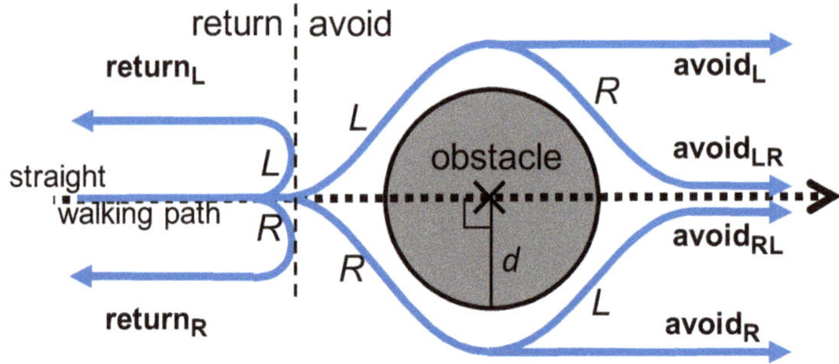

Figure 2. Definition of avoidance behavior.

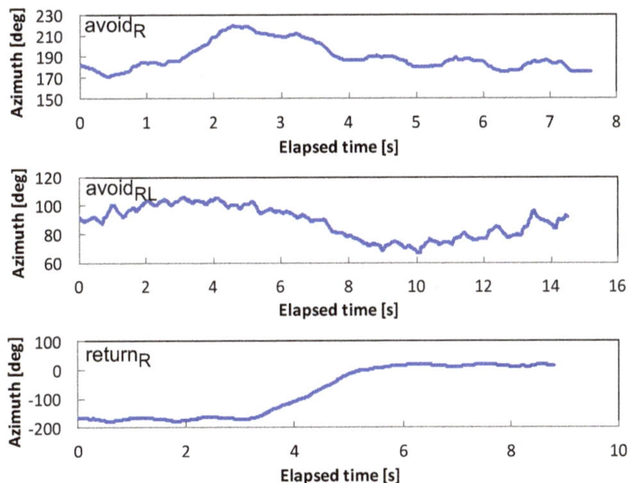

Figure 3. Raw azimuth signals of *avoid$_R$* (top), *avoid$_{RL}$* (center) and *return$_R$* (bottom).

3.3. Avoidance Behavior Recognition

As shown in Figure **??**, the recognition system takes a segment of the azimuth data as input and classifies the segment into one of six behaviors. The recognition task should be performed on streaming sensor data. Sliding variance can be calculated on streaming data to emphasize the start and the finish of the change of walking direction. However, a change of walking direction also occurs when a pedestrian turns a corner or walks along a curved road, which are normal behaviors and should not be detected as obstacle avoidance. Therefore, special care is required to distinguish these situations from one another. Automatic segmentation is beyond the focus of this article, and we utilize manually-segmented data to focus on recognizing the six behaviors.

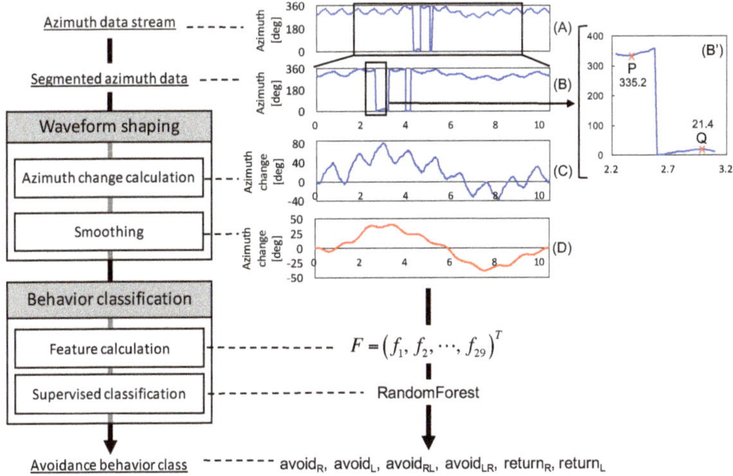

$$F = \left(f_1, f_2, \cdots, f_{29} \right)^T$$

Figure 4. Avoidance behavior recognition flow.

3.3.1. Waveform Shaping

The waveform shaping process works as a preprocess and is composed of azimuth change detection and smoothing. We obtain the azimuth value from an Android API. The value, which is calculated from accelerometer and magnetometer data, ranges from $0°$ to $359°$. Therefore, a non-contiguous change appears when the walking direction crosses the north, i.e., $0°$, as observed in B′ of Figure **??**. In this case, a person is supposed to change the direction from near west-northwest (P = $335.2°$) to near east-northeast (Q = $21.4°$). Furthermore, the value is normalized by converting it from the first value into a relative value. Then, the azimuth change in a segment is calculated by Algorithm **??**, in which Lines 7 to 9 handle a non-contiguous change. The transformed signal is shown in C of Figure **??**.

Algorithm 1 Calculate Azimuth Change Relative to the First Value in a Segment.

1: **procedure** AzimuthChange(a_{raw})   ▷ a_{raw} represents an array of a segment.
2:      $th \leftarrow 200$ ▷ A threshold value to judge if a non-contiguous change appears
3:      $segnum \leftarrow a_{raw}.length$
4:      $a_{change,0} \leftarrow 0$
5:      **for** $i \leftarrow 1, segnum - 1$ **do**
6:          $\delta \leftarrow a_{raw,i} - a_{raw,i-1}$
7:          **if** $|\delta| > th$ **then**
8:              $\delta \leftarrow \delta - sgn(\delta) \times 360$ ▷ $sgn(x)$ returns -1 if $x < 0$, 1 if $x > 0$, and 0 if $x = 0$.
9:          **end if**
10:          $a_{change,i} \leftarrow a_{change,i-1} + \delta$
11:      **end for**
12:      **return** a_{change} ▷ Returns an array of azimuth change relative to the initial value
13: **end procedure**

In addition, to remove the effect of body motion, i.e., smoothing, a moving average is applied as a low-pass filter, as shown in D of Figure **??**. The window size for the moving averages is 1/6 of the segment of azimuth data, as determined in a preliminary experiment.

3.3.2. Behavior Classification

Behavior classification, which consists of feature calculation and supervised classification, is performed after waveform shaping. In total, we specified 29 features, which are summarized in Table **??**. These features mainly contain basic statistics, such as mean, maximum, minimum, range, first and third quartiles, inter-quartile range (IQR), variance, standard deviation, summation, summation of squares, root mean square (RMS) and absolute values. In lining up features, we paid special attention to the fact that the trajectories of $avoid_{RL}$ and $avoid_R$ are clearly distinguished from each other after passing an obstacle (Figure **??**), so we split a segment into two parts at the center of the segment. Features calculated from the first half segment and the second half segment have subscripts FH and SH, respectively. In contrast, features from an entire segment have the subscript ALL. Note that Table **??** is listed in order of contribution to the classification. Not all of the corresponding effects of features can be significant or could even have a negative impact, which is discussed in Section **??**.

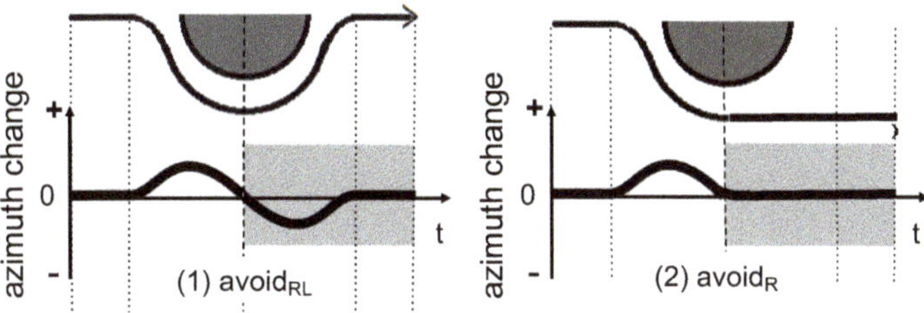

Figure 5. Segmentation of *avoid$_{RL}$* and *avoid$_R$*.

Table 1. Features, listed in order of contribution from upper left to lower right.

f_1	*min$_{SH}$*	f_{16}	*max$_{FH}$*
f_2	*max$_{SH}$*	f_{17}	*sum of squares$_{SH}$*
f_3	*mean$_{SH}$*	f_{18}	*variance$_{ALL}$*
f_4	*sum$_{SH}$*	f_{19}	*standard deviation$_{ALL}$*
f_5	*mean$_{FH}$*	f_{20}	*sum of absolute difference of*
f_6	*mean$_{ALL}$*		*both ends of 10 subsegments*
f_7	*median$_{ALL}$*	f_{21}	*sum$_{FH}$*
f_8	*ratio of range$_{ALL}$ to ΔSF*	f_{22}	*IQR$_{ALL}$*
f_9	*sum$_{ALL}$*	f_{23}	*3rd quartile$_{ALL}$*
f_{10}	*range$_{ALL}$*	f_{24}	*1st quartile$_{ALL}$*
f_{11}	*min$_{ALL}$*	f_{25}	*RMS$_{ALL}$*
f_{12}	*absolute min$_{ALL}$*	f_{26}	*sum of squares$_{ALL}$*
f_{13}	*max$_{ALL}$*	f_{27}	*variance$_{FH}$*
f_{14}	*absolute max$_{ALL}$*	f_{28}	*sum of squares$_{FH}$*
f_{15}	*min$_{FH}$*	f_{29}	*variance$_{SH}$*

ALL: an entire segment; FH and SH: first half and second half of a segment; IQR: inter-quartile range; RMS: root mean square; ΔSF: absolute difference between values at the start and the finish.

To calculate the eighth feature, (f_8) is introduced to represent the monotonicity of the azimuth change in a segment. As shown in Figure ??a and expressed by Equation (??), the feature gets larger as the maximum (*max$_{ALL}$*) and the minimum (*min$_{ALL}$*) values approach both ends (*start$_{ALL}$* and *finish$_{ALL}$*). We consider that "*return$_R$*" has the largest value of the three behaviors because the azimuth change of "*return$_R$*" is ideally a monotonic increase or decrease (see Figure ??).

$$f_8 = \left| \frac{max_{ALL} - min_{ALL}}{start_{ALL} - finish_{ALL}} \right| \tag{1}$$

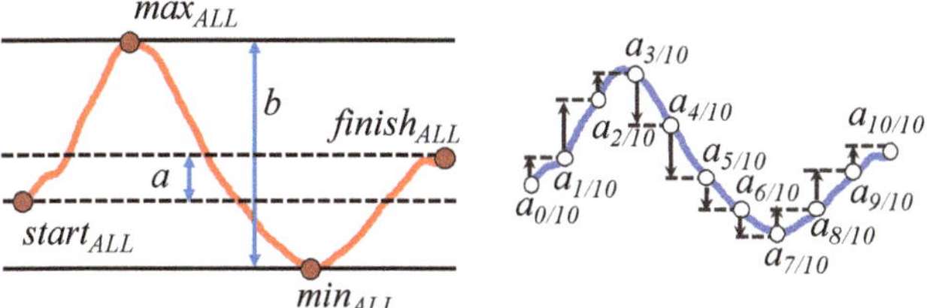

(a) Max-Min range and Start-Finish range for f_8

(b) Absolute difference of both ends of 10 subsegments for f_{20}

Figure 6. Notations for f_8 and f_{20}.

For the 20th feature (f_{20}), a segment is equally divided into 10 subsegments, and the absolute difference between both ends of each subsegment is summed up to 10 subsegments. The notation is illustrated in Figure **??**b, as well as expressed by Equation (**??**). The rationale for introducing this feature is that a behavior with a large azimuth change tends to have a large absolute difference. As shown in Figure **??**, $avoid_R$ has a smaller value than $avoid_{RL}$ due to a lack of azimuth change in the second half of the segment. Meanwhile, $return_R$ should have the largest change because the walking direction changes to the opposite side.

$$f_{20} = \sum_{i=0}^{9} |a_{(i+1)/10} - a_{i/10}| \tag{2}$$

Regarding the recognition (classification) functionality, handmade rule-based approaches [? ?], a statistical machine learning approach (J48decision tree) [?] and a probabilistic approach (hidden Markov model (HMM)) [?] were utilized in the literature of horizontal displacement recognition. A handmade rule-based approach can be considered as a form of decision tree in that "if-then" rules are set using the expert knowledge. Therefore, the decision-making process is more interpretive than what the J48 decision tree provides; however, the approach requires the careful design of the rule, and thus, the case with a small number of recognition classes seems to be suitable, e.g., three for [?] and two for [?]. The avoidance behavior recognition problem can be regarded as a time series pattern recognition, in which HMM is one of the following: popular approach speech [?], hand-written character [?] and gesture recognition [?]. However, the HMM-based system is considered to require a sizable amount of training to data to perform well [? ?]. Based on these considerations, we utilized a supervised learning classifier. The comparison among various types of classifiers is presented in Section **??**.

4. Offline Experiment

An offline experiment was carried out on various aspects, such as contributing features and the difference in individuals, the storing positions and the sizes of obstacles, in addition to the basic classification performance.

4.1. Dataset

Data collection was performed as summarized in Table **??**, and Figure **??** shows a scene of data collection. A "cross" mark was placed on the ground as an obstacle. Subjects were asked to walk on a straight path while avoiding obstacles with directed types of avoidance behaviors. They started walking about seven meters behind the center of the obstacle. The timing of the start and the finish in each avoidance behavior was based on their decisions, although they were asked to walk past a mark that represented the edge of an obstacle. The segmentation was done by hand.

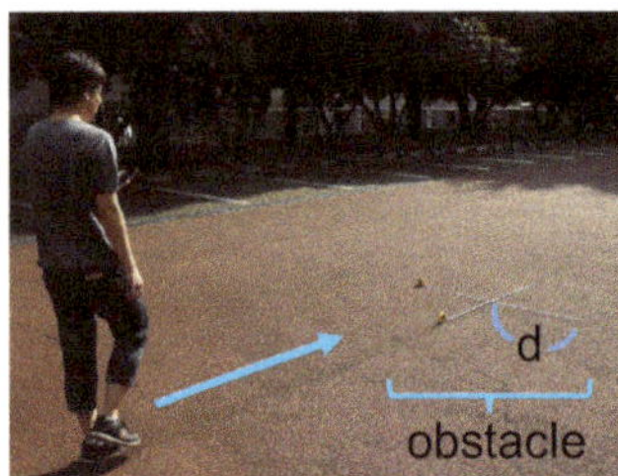

Figure 7. A scene of data collection.

Table 2. Condition of data collection.

Types of avoidance	$avoid_{RL}$, $avoid_R$, $return_R$
Size of obstacles (d)	0.2, 0.5, 0.7, 1.0, 1.5 m
Storing positions	hand (texting), trousers front pocket, trousers back pocket, chest pocket
Subjects	7 males and 2 females in their 20s
Number of trials	6 times per condition
Terminal	Samsung, Galaxy Nexus
Android version	Android 4.2.1
Sensor type	`Sensor.TYPE_ORIENTATION`
Sampling rate	10 Hz

In addition to the original data, we synthesized $avoid_{LR}$, $avoid_L$ and $return_L$ based on the findings that avoidance behaviors have left-right symmetry [?]. As shown by Equation (**??**), the synthesis is realized by inverting the sign of each sample. Here, $a_{L,k}$ and $a_{R,k}$ indicate the k-th sample in the collected data and in the synthesized data, respectively. Finally, the profile of the collected and synthesized dataset is summarized in Table **??**. Note that data with obstacle sizes 0.5 and 1.0 m were only used in Section **??** to evaluate the robustness of the classifier against the unknown size of obstacles.

$$a_{L,k} = -a_{R,k} \tag{3}$$

Table 3. Profile of the dataset.

Type	Segments	Person	Segments	Size	Segments
$avoid_{RL}$	865	A	540	0.2	844
$avoid_R$	866	B	262	0.5	486
$avoid_{LR}$	865	C	508	0.7	838
$avoid_L$	866	D	528	1.0	480
$return_R$	215	E	288	1.5	814
$return_L$	215	F	336	**Stored**	**Segments**
		G	358	hand	952
		H	538	trousers front pocket	966
		I	534	trousers back pocket	968
				chest pocket	1006

Note that the azimuth measurement relies on the magnetometer that may be affected by architectural construction, including metal, high-voltage current and magnetism. The data collection was carried out in an environment where no building and machinery exist around the subjects, in which we did not observe any unstable reading from the sensor. However, to observe if any disturbance in sensor reading exists, we empirically walked near air conditioner's outdoor units, vehicles, vending machines, exterior wall of buildings, etc., and visually checked a graph of the data stream. As a result, we found disturbance in very limited cases of passing by an electric vehicle and passing through a narrow passage surrounded by a reinforced concrete wall. In both cases, the data appear to be randomly and rapidly changing. Therefore, we consider that such a situation is distinguishable to avoid misrecognition of avoidance behavior; however, further study is required to recognize an avoidance behavior that occurs in such a situation.

4.2. Basic Classification Performance

4.2.1. Method

First of all, various types of classification methods, i.e., classifiers, are compared by applying 10-fold cross-validation (CV) to fix one classifier for later evaluation. Here, naive Bayes (a baseline approach), Bayesian network (a graphical model approach), multi-layer perceptron (MLP, an artificial neural network approach), sequential minimal optimization (SMO, a support vector machine approach), decision tree (J48) and random forest (an ensemble learning approach) were used. Table **??** summaries the parameters for each classifier used in a machine learning toolkit Weka ver. 7.3.13 [?], in which default values were utilized. The parameter symbols can be referred to as the reference manual of Weka.

Table 4. Classifier parameters in Weka.

Classifier	Parameter
Naive Bayes	N/A
Bayesian network	-Q K2 "-P 1 -S BAYES" -E SimpleEstimator "-A 0.5"
MLP	-L 0.3 -M 0.2 -N 500 -V 0 -E 20 -H a
SMO	-C 1.0 -P 1.0E-12 -K "PolyKernel -E 1.0 -C 250007"
J48	-C 0.25 -M 2
Random forest	-I 100 -K 0

4.2.2. Result and Analysis

Figure **??** shows the F-measure of each classifier, in which random forest performed the best followed by MLP. Another advantage of random forest is the small number of tuning parameters. In the Weka implementation, the number of major parameters is two, while that of MLP is five. Therefore, we consider that random forest is easy for tuning. Hereinafter, random forest with 100 trees is utilized.

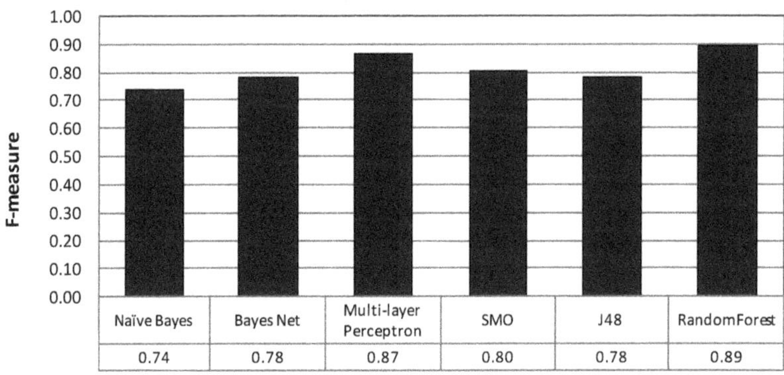

Figure 8. Comparison of various supervised classifier models.

Table **??** shows the confusion matrix, in which the row indicates the labeled class and the column is the recognition result. Table **??** summarizes the results of recall, precision and F-measure. Note that we normalized the recognition results by the smallest number of segments throughout this article because the number of data varies by class, as shown in Table **??**. The result shows an average classification performance with an F-measure of 0.89 and a range from 0.84 to 1.00.

Table 5. Confusion matrix of 10-fold cross-validation (CV) using random forest.

Label\Recognition	(1)	(2)	(3)	(4)	(5)	(6)
(1) $avoid_{RL}$	182	19	7	7	0	0
(2) $avoid_R$	25	179	6	6	0	0
(3) $avoid_{LR}$	7	6	183	19	0	0
(4) $avoid_L$	7	6	24	179	0	0
(5) $return_R$	0	0	0	0	215	0
(6) $return_L$	0	0	0	0	0	215

Table 6. Recall, precision and F-measure using random forest.

Class	Recall	Precision	F-Measure
$avoid_{RL}$	0.85	0.83	0.84
$avoid_R$	0.83	0.85	0.84
$avoid_{LR}$	0.85	0.83	0.84
$avoid_L$	0.83	0.85	0.84
$return_R$	1.00	1.00	1.00
$return_L$	1.00	1.00	1.00
Average	0.89	0.89	0.89

Classes $avoid_{RL}$ and $avoid_{LR}$ were misclassified into $avoid_R$ and $avoid_L$, respectively. We consider that this occurred because the second half of the segment of these classes is flat, which made it difficult to distinguish the classes from each other. In contrast, $return_R$ and $return_L$ were perfectly classified. We assumed that the start and the end of the walking direction of $avoid$ are identical, whereas the start and the end of $return$ are different, i.e., on the opposite side. We consider that the features of $return$ had large differences from those of $avoid$.

The recognition of road anomaly class is useful for a road administration office; however, to prioritize their repairing tasks, the size of the road anomaly should be recognized, since a large avoidance behavior indicates the significance of the anomaly. Currently, we have a dataset with five obstacle sizes. Defining new classes for each size is not practical. Therefore, we will build a regression model based on some features and the obstacle size, which will be applied after classifying datasets into the six avoidance behavior classes.

4.3. Feature Relevance

4.3.1. Method

To understand effective features for avoidance behavior recognition, the relevance of features was evaluated based on information theory. Information gain is commonly used in feature selection, where the gain of information provided by a particular feature is calculated by subtracting a conditional entropy with that feature from the entropy under a random guess [?]. We used InfoGainAttributeEval and Ranker in Weka [?] as implementations for evaluating information gain and generating ranking, respectively.

As described in Section **??**, Table **??** is already listed in order of contribution (relevance) with the above implementations. To observe the change of classification performance against the number of features, a 10-fold cross-validation was carried out against a dataset with the top-k features, and F-measures were calculated. Here, k varies from one (best) to 29 (all).

4.3.2. Result and Analysis

As shown in Figure **??**, the F-measure for the top five features rapidly increases. From the top six to 20 features, the F-measure gradually increases by very slight up-and-down movements. Finally, the increase almost levels off for more than 20 features. As described in Section **??**, we split a segment into two parts: first half (FH) and second half (SH) (see Figure **??**). By looking at Table **??**, the top-four contributing features are derived from SH. This indicates that the decision to calculate features by dividing them into FH and SH proved to be correct. Note that the division of the two parts is based on the number of samples in a segment under the assumption that people walk at a constant velocity; however, in practice, the speed might change in the vicinity of an obstacle. This makes the two divided parts not as clear as those shown in Figure **??**. Detecting such a changing point would emphasize the difference of activities more clearly and improve the performance.

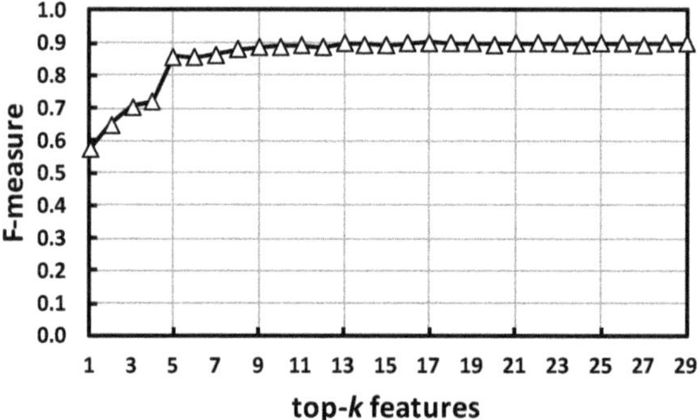

Figure 9. Classification performance with top-k contributing features. Features are provided in Table **??**.

Furthermore, the "ratio of range$_{ALL}$ to ΔSF" (f_8), which was added to get the the monotonicity of the change, appeared as the eighth in the ranking. In the two-thirds of the ranking, features that eliminate negative values are found (f_{12}, f_{14}, f_{17}, f_{25}, f_{26} and f_{28}). This implies that the component of direction is more important than the magnitude of movement. As opposed to the expectation on the accumulated azimuth change described in Section **??**, "sum of the absolute difference of both ends of 10 subsegments" appears in the 20th ranking (f_{20}). We consider that f_{20} contributes to discriminate *return* from *avoid* because of the large change in the direction. However, the difference of the accumulated change between $avoid_R$ and $avoid_L$ is not so large as other features and less contributive. Furthermore, since f_{20} represents the amount of change, it is difficult to discriminate $avoid_{RL}$ from $avoid_{LR}$, which are symmetric about the horizontal axis.

4.4. Person Dependency

4.4.1. Method

The classification performance shown in Section **??** was obtained by 10-fold cross-validation against the dataset from all subjects. This represents the average performance of the classification method. To evaluate the method under realistic conditions, where the data of a user are not used to train the classifier, we conducted leave-one-subject-out cross-validation (LOSO-CV). In LOSO-CV, the dataset of a particular subject is used for testing purposes, while the dataset of the rest of the subject group, i.e., eight subjects, is utilized for training a classifier. This process was iterated for all subjects, and an average was calculated.

It is not difficult to foresee that the best performance comes from using a personalized classifier, in which a classifier is trained with the dataset of a particular person and tested with the dataset of the same person (e.g., [?]). Therefore, to see the performance under this best condition, we conducted 10-fold cross-validation using personalized classifiers. This evaluation is referred to as self-CV.

4.4.2. Result and Analysis

Figure **??** shows the comparison in individual differences (Subjects A to I) and the averages for the two evaluation conditions. As shown in this figure, all subjects had equal or better classification performance with self-CV than with LOSO-CV. On average, self-CV performed better than LOSO-CV with an F-measure difference of 0.07.

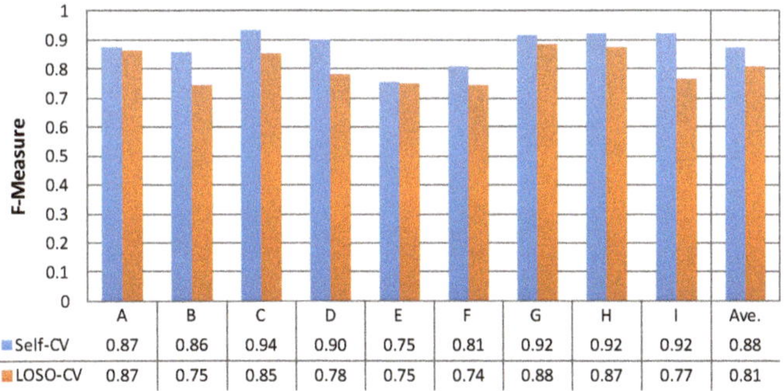

	A	B	C	D	E	F	G	H	I	Ave.
Self-CV	0.87	0.86	0.94	0.90	0.75	0.81	0.92	0.92	0.92	0.88
LOSO-CV	0.87	0.75	0.85	0.78	0.75	0.74	0.88	0.87	0.77	0.81

Figure 10. Individual differences in self-cross-validation (CV) and leave-one-subject-out cross-validation (LOSO-CV).

The differences between the F-measure of self-CV and LOSO-CV for Subjects B, D and I are relatively large (0.11, 0.12 and 0.15, respectively). To analyze the reasons, a classifier was tuned for each subject. The classifiers were then tested with the datasets from other subjects. Figure **??** summarizes the resulting F-measures. In this table, the grayscale levels are normalized between the minimum (0.54) and the maximum (0.94) values to white and black, respectively. Furthermore, the values on the diagonal line of the same subject IDs indicate the F-measures of self-CV, shown also in the first row of Figure **??**. As a result, we consider that "noisy" data are included in the LOSO-CV for Subjects B, D and I. In other words, some of the training data may have some subjects whose data were incompatible with Subjects B, D and I. In Figure **??**, Subject H seems to be incompatible with Subjects B and D, as shown by the lowest values in each column (0.56 and 0.54, respectively). Furthermore, Subject B seems to be an incompatible subject for Subject I (0.58). To validate these thoughts, we built classifiers with datasets from "compatible" subjects as follows. First, an average of the F-measure excluding the self-CV value is calculated for each column. Then, the datasets of the subjects whose personalized classifiers performed better than the average are used to train a new classifier. Hence, Subjects C, D, E, F and G were selected as compatible subjects for Subject B, while Subjects A, B, E, F and G were selected as compatible subjects for Subject D. Similarly, as compatible subjects for Subject I, Subjects A, E, G and H were selected. As a result of testing with these new classifiers, the F-measures of LOSO-CV against Subjects B, D and I were improved to 0.80, 0.85 and 0.86, respectively; this was an increase of 0.05, 0.07 and 0.09 from the original LOSO-CV. In the future, we will investigate a method to find compatible persons to build a classifier in a systematic manner.

The F-measures of self-CV for Subjects E and F are relatively low, i.e., 0.75 and 0.81, respectively. This indicates that the features obtained from them failed to capture the characteristics of the target behaviors due to the large variation within subjects. Moreover, the average number of segments per class for Subjects E and F are 48 and 52, respectively, as calculated from Table **??**. Therefore, the classifiers for these subjects are trained with around 44 and 47 segments (nine-tenths of the number of segments), respectively. We consider that these classifiers were not trained sufficiently.

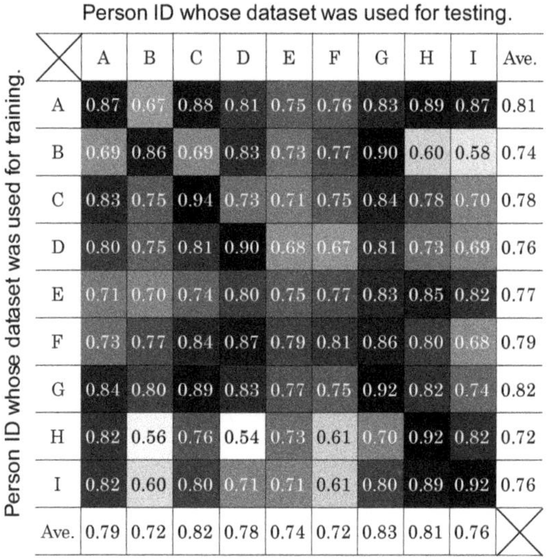

Person ID whose dataset was used for testing.

Person ID whose dataset was used for training.	A	B	C	D	E	F	G	H	I	Ave.
A	0.87	0.67	0.88	0.81	0.75	0.76	0.83	0.89	0.87	0.81
B	0.69	0.86	0.69	0.83	0.73	0.77	0.90	0.60	0.58	0.74
C	0.83	0.75	0.94	0.73	0.71	0.75	0.84	0.78	0.70	0.78
D	0.80	0.75	0.81	0.90	0.68	0.67	0.81	0.73	0.69	0.76
E	0.71	0.70	0.74	0.80	0.75	0.77	0.83	0.85	0.82	0.77
F	0.73	0.77	0.84	0.87	0.79	0.81	0.86	0.80	0.68	0.79
G	0.84	0.80	0.89	0.83	0.77	0.75	0.92	0.82	0.74	0.82
H	0.82	0.56	0.76	0.54	0.73	0.61	0.70	0.92	0.82	0.72
I	0.82	0.60	0.80	0.71	0.71	0.61	0.80	0.89	0.92	0.76
Ave.	0.79	0.72	0.82	0.78	0.74	0.72	0.83	0.81	0.76	

Figure 11. Applicability of person-dependent classification model.

4.5. Effect of Sensor Storing Position

4.5.1. Method

As investigated by Ichikawa et al. [?], people carry their smartphone terminals in various positions, such as their trousers pocket and chest pocket. We carried out an experiment to see the impact of storing position on the classification performance. The experiment was carried out by training a classifier with a dataset from a particular position and testing the classification with the datasets from the other positions.

4.5.2. Result and Analysis

Table **??** summarizes the F-measure results. In this table, the row indicates the storing positions from which datasets for training position-specific classifiers were obtained, and the column represents the datasets for testing. Note that the values on the diagonal line at the same positions were obtained by 10-fold cross-validation. These values indicate the ideal performance when classifiers are tuned for dedicated positions, and the average is 0.87. The table demonstrates that the classifiers tuned for particular positions did not predominantly perform best. This observation allows us to propose two approaches for constructing classification models.

Table 7. Robustness to storing position variation.

Trained with\Test with	(1)	(2)	(3)	(4)	Average
(1) Hand (texting)	0.86	0.82	0.82	0.91	0.86
(2) Trousers front pocket	0.85	0.85	0.86	0.89	0.86
(3) Trousers back pocket	0.83	0.85	0.88	0.88	0.86
(4) Chest pocket	0.86	0.83	0.80	0.91	0.85
Average	0.85	0.84	0.84	0.90	–

The first approach is a straightforward one that constructs a single classifier with the datasets from all positions. This is the case shown in Section **??**, and we obtain an F-measure of 0.89. This is better than the average of the tuned classifier approach (0.87). However, to realize this approach, datasets from all positions should be collected. The second approach is to share classifiers with some positions. In this case, a tuned classifier for the position "hand" is shared with the case in which the position of a terminal is judged as "chest pocket", since the performance of "chest pocket" using a classifier tuned for "hand" is as high as with a tuned classifier for "chest pocket". This can omit training data collection for the "chest pocket" classifier. Similarly, a classifier built from the dataset from "trousers back pocket" is shared with the data obtained from "trousers front pocket". As shown in [?], front and back trousers pockets are often misrecognized for each other. Therefore, sharing the classifier between two positions can become robust against the mistake of the underlying storing position recognizer. In the second approach, the averaged F-measure is 0.88. Table **??** summarizes the result. The second approach has a slightly worse F-measure; however, it just needs to collect the dataset from two positions, "hand" and "trousers back pocket", which we consider a great advantage in reducing the cost of data collection. Such low-cost modeling accelerates the deployment of the system. The sharing approach may sacrifice the accuracy of recognition; however, it could be improved on the server side if a number of people utilized the system.

Table 8. Dealing with position dependency.

Approach	Average
(0) Tuned classifier for each position	0.87
(1) Single classifier with the dataset from all positions	0.89
(2) Sharing classifiers with some positions	0.88

4.6. Robustness to Unknown Obstacle Size

4.6.1. Method

The performance evaluations above were performed by the classifiers trained by datasets with the obstacle sizes 0.2, 0.7 and 1.5 m. To understand the robustness against unknown sizes of obstacles, we used the datasets of obstacle sizes 0.5 and 1.0 m for the test, in which the dataset of obstacle sizes 0.2, 0.7 and 1.5 m was used to train the classifiers, as before.

4.6.2. Result and Analysis

The F-measures of the results are shown in Table **??**, where we can find that all values are better than the ones in the rightmost column in Table **??**. We consider that this is because the obstacle size used for this test is in the range of the training dataset, i.e., 0.2 to 1.5 m. Therefore, the features obtained from the dataset with unknown obstacle sizes might fit into the ranges of trained features. The result implies that a classifier can be trained with a limited size of obstacles, i.e., probably for detecting upper, middle and lower sizes of obstacles.

Table 9. Performance against unknown obstacle size.

Class	$avoid_{RL}$	$avoid_R$	$avoid_{LR}$	$avoid_L$	$return_R$	$return_L$	Average
F-measure	0.93	0.91	0.92	0.91	1.00	1.00	0.94

5. Conclusions

In this article, we proposed a road anomaly detection system based on opportunistic sensing by using pedestrians' smartphone terminals. Opportunistic sensing requires no explicit user involvement, which is expected to lower the barrier of people's participation to the sensing activity. Although automatic road anomaly detection methods have already been proposed for cars and bikes, we considered that pedestrians' avoidance behaviors are too slight to adapt these existing methods. After showing the overall system concept, we focused on the design of an obstacle avoidance behavior recognition system, in which waveform shaping, feature extraction and supervised classifiers were presented as major components. Six classes of avoidance behaviors were defined to test the recognition system from various aspects after collecting data from nine people with 410 trials on average. The following results were obtained:

- A 10-fold CV showed an average classification performance with an F-measure of 0.89 for six avoidance behaviors.
- The recognition system could handle the obstacle sizes of 0.2 to 1.5 m. Untrained sizes of obstacle avoidance were also recognized with an F-measure of 0.94.
- A user-independent classifier classified six avoidance behaviors with an F-measure of 0.81. The possibility of improving a user-independent classification by choosing classifiers trained by compatible persons was shown.
- Features resulting from (1) splitting a segment into the first half and the second half and (2) considering the monotonicity of change effectively recognized avoidance behaviors.
- The performance slightly depends on the sensor (smartphone) storing position on the body. Selecting a classifier for a particular position improves the performance. To reduce the cost of data collection, only the data from "hand" and "trousers back pocket" need be collected.

The results are obtained under an ideal and controlled environment; however, the results indicate that the proposed recognition method is robust against the size of obstacles and that the dependency on the storing position of a smartphone can be handled by an appropriate classifier per storing position. Furthermore, an analysis implies that classification of data from an "unknown" person can be improved by taking into account the compatibility of a classifier. The next step toward an all-in-one road anomaly detection system is to investigate an automatic avoidance event segmentation method that was performed by hand in this article. The key challenge is to discriminate normal behaviors that are associated with a change of walking direction, e.g., a pedestrian turns a corner or walks along a curved road, from true obstacle avoidance. We will leverage the characteristics in the azimuth difference and the walking distance to complete the change of walking direction to distinguish these situations. A real-world experiment is also required to assess the robustness of the proposed system. Lower-power operation is a critical issue for opportunistic sensing to be accepted by people because GPS-based positioning is generally a power-intensive approach [?]. Unlike continuous positioning, such as a noise map [?], our system can take an event-driven positioning, in which the positioning is performed only when an obstacle is detected. The challenge here is the positioning error due to the delay of activating a GPS receiver, i.e., the actual position of an avoidance event may be backward from the position where a GPS receiver returns. We will investigate a correction method by leveraging a pedestrian dead-reckoning (PDR) technology. Finally, server-side aggregation and the filtering technique will be investigated to realize the overall system.

Acknowledgments: This work was supported by an operational grant from Tokyo University of Agriculture and Technology and a grant-in-aid from Foundation for the Fusion of Science and Technology.

Author Contributions: Both authors collaboratively worked on this article in different ways that range from conceptualization and design of the recognition model, to the conduct of the study and the editing of the article. Both authors contributed equally enough to warrant their co-authorship.

Conflicts of Interest: The authors declare no conflict of interest.

References

.   City of Chiba. Chiba-Repo Field Trial: Review Report. 2013. Available online: http://www.city.chiba.jp/shimin/shimin/kohokocho/documents/chibarepo-hyoukasho.pdf (accessed on 30 September 2016).

.   mySociety Limited. FixMyStreet. Available online: http://fixmystreet.org (accessed on 30 September 2016).

. Goldman, J.; Shilton, K.; Burke, J.; Estrin, D.; Hansen, M.; Ramanathan, N.; Reddy, S.; Samanta, V.; Srivastava, M.; West, R. *Participatory Sensing: A Citizen-Powered Approach to Illuminating the Patterns that Shape Our World*; Foresight and Governance Project, White Paper; Woodrow Wilson International Center for Scholars: Washington, DC, USA, 2009.

. Lane, N.D.; Miluzzo, E.; Lu, H.; Peebles, D.; Choudhury, T.; Campbell, A.T. A survey of mobile phone sensing. *IEEE Commun. Mag.* **2010**, *48*, 140–150.

. Carrera, F.; Guerin, S.; Thorp, J.B. By the people, for the people: The crowdsourcing of "STREETBUMP": An automatic pothole mapping app. *ISPRS Int. Arch. Photogramm. Remote Sens. Spat. Inf. Sci.* **2013**, *XL-4/W1*, 19–23.

. Chen, D.; Cho, K.T.; Han, S.; Jin, Z.; Shin, K.G. Invisible Sensing of Vehicle Steering with Smartphones. In Proceedings of the 13th Annual International Conference on Mobile Systems, Applications, and Services (MobiSys '15), Florence, Italy, 18–22 May 2015; pp. 1–13.

. Eriksson, J.; Girod, L.; Hull, B.; Newton, R.; Madden, S.; Balakrishnan, H. The Pothole Patrol: Using a Mobile Sensor Network for Road Surface Monitoring. In Proceedings of the 6th International Conference on Mobile Systems, Applications, and Services (MobiSys '08), Breckenridge, CO, USA, 17–20 June 2008; pp. 29–39.

. Kaneda, S.; Asada, S.; Yamamoto, A.; Kawachi, Y.; Tabata, Y. A Hazard Detection Method for Bicycles by Using Probe Bicycle. In Proceedings of the IEEE 38th International Computer Software and Applications Conference Workshops (COMPSACW), Västerås, Sweden, 21–25 July 2014; pp. 547–551.

. Mohan, P.; Padmanabhan, V.N.; Ramjee, R. Nericell: Rich Monitoring of Road and Traffic Conditions Using Mobile Smartphones. In Proceedings of the 6th ACM Conference on Embedded Network Sensor Systems (SenSys '08), Raleigh, NC, USA , 4–7 November 2008; pp. 323–336.

. Ishikawa, T.; Fujinami, K. Pedestrian's Avoidance Behavior Recognition for Road Anomaly Detection in the City. In Proceedings of the ACM International Joint Conference on Pervasive and Ubiquitous Computing and ACM International Symposium on Wearable Computers (UbiComp/ISWC '15), Osaka, Japan, 7–11 September 2015; pp. 201–204.

. Bhoraskar, R.; Vankadhara, N.; Raman, B.; Kulkarni, P. Wolverine: Traffic and road condition estimation using smartphone sensors. In Proceedings of the IEEE Fourth International Conference onCommunication Systems and Networks (COMSNETS), Bangalore, India, 3–7 January 2012; pp. 1–6.

. Kamimura, T.; Kitani, T.; Kovacs, D.L. Automatic classification of motorcycle motion sensing data. In Proceedings of the IEEE International Conference on Consumer Electronics—Taiwan (ICCE-TW), Taipei, Taiwan, 26–28 May 2014; pp. 145–146.

. Seraj, F.; van der Zwaag, B.J.; Dilo, A.; Luarasi, T.; Havinga, P.J.M. RoADS: A road pavement monitoring system for anomaly detection using smart phones. In Proceedings of the 1st International Workshop on Machine Learning for Urban Sensor Data, SenseML 2014, Nancy, France, 15 September 2014; pp. 1–16.

. Thepvilojanapong, N.; Sugo, K.; Namiki, Y.; Tobe, Y. Recognizing bicycling states with HMM based on accelerometer and magnetometer data. In Proceedings of the SICE Annual Conference (SICE), Tokyo, Japan, 13–18 September 2011; pp. 831–832.

. Iwasaki, J.; Yamamoto, A.; Kaneda, S. Road information-sharing system for bicycle users using smartphones. In Proceedings of the IEEE 4th Global Conference on Consumer Electronics (GCCE), Osaka, Japan, 27–30 October 2015; pp. 674–678.

. Jain, S.; Borgiattino, C.; Ren, Y.; Gruteser, M.; Chen, Y.; Chiasserini, C.F. LookUp: Enabling Pedestrian Safety Services via Shoe Sensing. In Proceedings of the 13th Annual International Conference on Mobile Systems, Applications, and Services (MobiSys '15), Florence, Italy, 18–22 May 2015; pp. 257–271.

. Alessandroni, G.; Klopfenstein, L.C.; Delpriori, S.; Dromedari, M.; Luchetti, G.; Paolini, B.D.; Seraghiti, A.; Lattanzi, E.; Freschi, V.; Carini, A.; et al. SmartRoadSense: Collaborative Road Surface Condition Monitoring. In Proceedings of the 8th International Conference on Mobile Ubiquitous Computing, Systems, Servies and Technologies, Rome, Italy, 24–28 August 2014; pp. 210–215.

. Tatebe, K.; Nakajima, H. Avoidance behavior against a stationary obstacle under single walking: A study on pedestrian behavior of avoiding obstacles (I). *J. Archit. Plan. Environ. Eng.* **1990**, *418*, 51–57.

. Rabiner, L.R. A tutorial on hidden Markov models and selected applications in speech recognition. *Proc. IEEE* **1989**, *77*, 257–286.

ISPRS Int. J. Geo-Inf. **2016**, *5*, 182

- Hu, J.; Brown, M.K.; Turin, W. HMM based online handwriting recognition. *IEEE Trans. Pattern Anal. Mach. Intell.* **1996**, *18*, 1039–1045.
- Wilson, A.D.; Bobick, A.F. Parametric Hidden Markov Models for gesture recognition. *IEEE trans. Pattern Anal. Mach. Intell.* **1999**, *21*, 884–900.
- Liu, J.; Zhonga, L.; Wickramasuriyab, J.; Vasudevanb, V. uWave: Accelerometer-based personalized gesture recognition , its applications. *Pervasive Mob. Comput.* **2009**, *5*, 657–675.
- Rajko, S.; Qian, G.; Ingalls, T.; James, J. Real-time gesture recognition with minimal training requirements and on-line learning. In Proceedings of the IEEE Conference on Computer Vision and Pattern Recognition, Minneapolis, MN, USA, 17–22 June 2007; pp. 1–8.
- Machine Learning Group at University of Waikato. Weka 3—Data Mining with Open Source Machine Learning Software in Java. Available online: http://www.cs.waikato.ac.nz/ml/weka/ (accessed on 30 September 2016).
- Witten, I.H.; Frank, E.; Hall, M.A. *Data Mining: Practical Machine Learning Tools and Techniques*, 3rd ed.; Morgan Kaufmann Publishers: San Francisco, CA, USA, 2011.
- Fujinami, K.; Kouchi, S. Recognizing a Mobile Phone's Storing Position as a Context of a Device and a User. In Proceedings of the 9th International Conference on Mobile and Ubiquitous Systems: Computing, Networking and Services (MobiQuitous), Beijing, China, 12–14 December 2012; pp. 76–88.
- Ichikawa, F.; Chipchase, J.; Grignani, R. Where's The Phone? A Study of Mobile Phone Location in Public Spaces. In Proceedings of the 2nd International Conference on Mobile Technology, Applications and Systems, Guangzhou, China, 15–17 November 2005; pp. 1–8.
- Ben Abdesslem, F.; Phillips, A.; Henderson, T. Less is More: Energy-efficient Mobile Sensing with Senseless. In Proceedings of the 1st ACM Workshop on Networking, Systems, and Applications for Mobile Handhelds (MobiHeld '09), Barcelona, Spain, 16–21 August 2009; pp. 61–62.
- Rana, R.K.; Chou, C.T.; Kanhere, S.S.; Bulusu, N.; Hu, W. Ear-phone: An End-to-end Participatory Urban Noise Mapping System. In Proceedings of the 9th ACM/IEEE International Conference on Information Processing in Sensor Networks (IPSN '10), Stockholm, Sweden, 12–15 April 2010; pp. 105–116.

 International Journal of
Geo-Information

Article

A High-Efficiency Method of Mobile Positioning Based on Commercial Vehicle Operation Data

Chi-Hua Chen [1,2,*], **Jia-Hong Lin** [1], **Ta-Sheng Kuan** [1] **and Kuen-Rong Lo** [1]

[1] Telecommunication Laboratories, Chunghwa Telecom Co., Ltd., Taoyuan 326, Taiwan; arphen@cht.com.tw (J.-H.L.); ditto@cht.com.tw (T.-S.K.); lo@cht.com.tw (K.-R.L.)

[2] Department of Information Management and Finance, National Chiao Tung University, Hsinchu 300, Taiwan

* Correspondence: chihua0826@gmail.com; Tel.: +886-3-4244091

Academic Editor: Wolfgang Kainz
Received: 26 March 2016; Accepted: 26 May 2016; Published: 2 June 2016

Abstract: Commercial vehicle operation (CVO) has been a popular application of intelligent transportation systems. Location determination and route tracing of an on-board unit (OBU) in a vehicle is an important capability for CVO. However, large location errors from global positioning system (GPS) receivers may occur in cities that shield GPS signals. Therefore, a highly efficient mobile positioning method is proposed based on the collection and analysis of the cellular network signals of CVO data. Parallel- and cloud-computing techniques are designed into the proposed method to quickly determine the location of an OBU for CVO. Furthermore, this study proposes analytical models to analyze the availability of the proposed mobile positioning method with various outlier filtering criteria. Experimentally, a CVO system was designed and implemented to collect CVO data from Chunghwa Telecom vehicles and to analyze the cellular network signals of CVO data for location determination. A case study found that the average errors of location determination using the proposed method *vs.* using the traditional cell-ID-based location method were 163.7 m and 521.2 m, respectively. Furthermore, the practical results show that the average location error and availability of using the proposed method are better than using GPS or the cell-ID-based location method for each road type, particularly urban roads. Therefore, this approach is feasible to determine OBU locations for improving CVO.

Keywords: mobile positioning; commercial vehicle operation data; cellular network; cloud computing

---

1. Introduction

In recent years, consumption and logistics patterns have changed with the development of economies and the gradual increases in national incomes. Furthermore, distribution channels have changed from direct distribution to the use of distribution centers (DCs). DCs can ship goods both upstream and downstream. The use of intelligent transportation fleet management technology to control the transit of goods is more efficient than traditional methods, and logistics companies receive higher economic benefits. Therefore, commercial vehicle operation (CVO) has been a popular application of intelligent transportation systems (ITS). The components of CVO include fleet administration, freight administration, electronic clearance, commercial vehicle administrative processes, international border crossing clearance, weigh-in-motion, roadside CVO safety, on-board safety monitoring, CVO fleet maintenance, hazardous material planning and incident response, freight in-transit monitoring, and freight terminal management.

Location determination of a vehicle by an on-board unit (OBU) is important for CVO. Precise location information can be used to support fleet administration, freight administration, freight in-transit monitoring, *etc.* For example, DCs can monitor the locations and movements of OBUs and provide the status of freight to users and receivers. The Global Positioning System (GPS) is the most popular location determination method for OBUs [1]. However, because of interference with GPS signals in cities, location errors may be generated from GPS. Therefore, this paper considers mobile positioning techniques to obtain location information when the signals of GPS satellites are weak. Various techniques have been proposed to analyze the signals from radio frequency identification (RFID) [2], Bluetooth [3], wireless local area networks (WLAN) [4–6], wireless sensor networks (WSN) [7,8], and cellular networks [9] for location determination [10]. However, the transmission ranges of RFID, Bluetooth, and WLAN are short, and they may be not suitable for CVO. Therefore, this study proposes a high-efficiency mobile positioning method to analyze the cellular network signals of CVO data. The method can be combined with cloud computing techniques to quickly determine the location of an OBU. Furthermore, a CVO system is proposed and implemented in this study, including OBUs and a CVO server.

The remainder of the paper is as follows. Section 2 presents and discusses the various techniques that exploit the cellular network for location determination. A high-efficiency mobile positioning method based on CVO data is proposed and illustrated in Section 3. Section 4 proposes analytical models to analyze the feasibility of the proposed mobile positioning method with various outlier filtering criteria. Section 5 describes a CVO system implementation and analyzes practical records to evaluate the proposed method. Finally, Section 6 discusses our conclusions and proposed future work.

2. Cellular-Based Positioning Methods

The 3rd Generation Partnership Project (3GPP) defined three classes of cellular-based positioning methods: the assisted global position system (A-GPS), mobile scan report (MSR)-based location methods, and database lookup methods [9].

2.1. Assisted GPS

A-GPS is designed to transfer almanac data from an assisted-positioning server to a mobile device through a network connection [11,12]. This method requires less time than traditional GPS method, approximately thirty seconds, for searching for satellites and determining location.

2.2. MSR-Based Location Methods

The MSRs, which include received signal strength indication (RSSI), round-trip delay (RTD), and relative delay (RD), are analyzed for location determination. MSR-based location methods can be classified into three categories: angle of arrival (AoA) [13–15], time of arrival (ToA) [13,14,16], and time difference of arrival (TDoA) [9,10,13]. This approach requires higher computation power than other methods [12,17–19].

2.3. Database Lookup Methods

Database lookup methods are used to determine the location of a mobile device quickly through static database queries. These methods can be classified into three categories: a cell-ID-based method [19], a handover-based method [20–22], and a fingerprint positioning method [23–28]. However, the lengths of cells and handoff zones are approximately 2 km and 200 m, respectively, and the location error depends on the cell size [12,19,29]. Although precise location estimation can be obtained by the fingerprint positioning method, higher computation power and an establishment fee are required [30].

To resolve these disadvantages, this study proposes a high-efficiency mobile positioning method based on the database lookup method to determine the location of mobile device quickly. Additionally, this method considers the RSSI of the connected cell to increase the accuracy of the estimated location.

3. High-Efficiency Mobile Positioning Method

A high-efficiency mobile positioning method is proposed to analyze and determine the location of each cell-RSSI pair from CVO data. The following subsections present two stages: (1) a pre-deployment stage and (2) a runtime stage.

3.1. Pre-Deployment Stage

In this stage, a mobile positioning algorithm is proposed to retrieve the location information and cellular network signals (*i.e.*, cell ID and RSSI) of historical data from GPS-equipped commercial vehicles and to estimate location for each cell-RSSI pair. For the computation requirements of the voluminous CVO data, the MapReduce programming model [31] and HIVE [32] are assumed to be built into cloud computing environments for quick location determination.

3.1.1. Input Data

Input data are CVO data. Each record of CVO data includes the longitude Hive and latitude of an OBU, the ID of the current connected cell, and the RSSI of the connected cell. Each OBU can periodically send CVO data (*i.e.*, longitude and latitude of the OBU, the ID of the current connected cell, and the RSSI of the connected cell) to CVO servers every 30 s. The RSSI and cell ID are paired and called the cell-RSSI pair. For example, the cell ID is presented as 10721_47366 when location area code (LAC) is 10721 and service area code (SAC) is 47366. The cell-RSSI pair is presented as 10721_47366_21 when the cell ID is 10721_47366 and the RSSI is −21 dBm [33].

The notations of this study are summarized below:

- There are n cells in CVO data, and the RSSI range of each cell is between 0 and m dBm. Therefore, there are $n \times m$ cell-RSSI pairs, and all cells have the same RSSI range.
- The number of records of the i-th cell-RSSI pair is defined as r_i.
- The longitude of the j-th record of the i-th cell-RSSI pair is defined as $x_{i,j}$, and the latitude of the j-th record of the i-th cell-RSSI is defined as $y_{i,j}$.
- The mean of $x_{i,j}$ is defined as $\mu_{x,i}$, and the mean of $y_{i,j}$ is defined as $\mu_{y,i}$.
- The standard deviation of $x_{i,j}$ is defined as $\sigma_{x,i}$, and the standard deviation of $y_{i,j}$ is defined as $\sigma_{y,i}$.
- After outlier filtering, the number of records of the i-th cell-RSSI pair is defined as r_i'.
- After outlier filtering, the longitude of the j-th record of the i-th cell-RSSI pair is defined as $x_{i,j}'$, and the latitude of the j-th record of the i-th cell-RSSI is defined as $y_{i,j}'$.
- The mean of $x_{i,j}'$ is defined as $\mu_{x,i}'$, and the mean of $y_{i,j}'$ is defined as $\mu_{y,i}'$.
- The longitude of the i-th cell-RSSI pair x_i is assumed to be normally distributed with mean $\mu_{x,i}$ and standard deviation $\sigma_{x,i}$. The probability density function (PDF) of its normal distribution is defined as $f\left(x_i, \mu_{x,i}, \sigma_{x,i}\right) = \frac{1}{\sigma_{x,i}\sqrt{2\pi}}e^{-\frac{(x_i-\mu_{x,i})^2}{2\sigma_{x,i}^2}}$ [33,34].
- The latitude of the i-th cell-RSSI pair y_i is assumed to be normally distributed with mean $\mu_{y,i}$ and standard deviation $\sigma_{y,i}$. The PDF of its normal distribution is defined as $f\left(y_i, \mu_{y,i}, \sigma_{y,i}\right) = \frac{1}{\sigma_{y,i}\sqrt{2\pi}}e^{-\frac{(y_i-\mu_{y,i})^2}{2\sigma_{y,i}^2}}$ [33,34].

A study of cell-RSSI pair 10721_47366_21 helps to evaluate the assumptions of longitude and latitude distributions. The historical location data of this cell-RSSI pair were collected by the CVO system of Chunghwa Telecom from November 2013 to January 2014. The cumulative distribution functions (CDFs) of longitudes and latitudes based on 13,231 historical records were calculated and are illustrated in Figure 1. The chi-square goodness of fit test [26,35,36] is used to evaluate the distributions of practical data and normal distribution. Chi-square tests of these assumptions showed that $\chi^2 = 0.653 < \chi^2_{11,0.05} = 19.675$ when $\alpha = 0.05$ for the longitude distribution and

$\chi^2 = 0.414 < \chi^2_{11,0.05} = 19.675$ when $\alpha = 0.05$ for the latitude distribution. No significant difference was observed, so the distributions of longitude and latitude were similar to normal distributions.

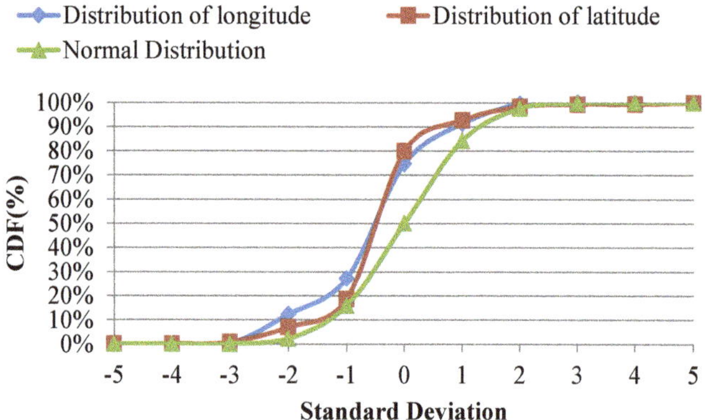

Figure 1. The distributions of longitudes and latitudes.

3.1.2. Process

Three steps comprise the mobile positioning method, as follows: (a) computation of the mean and standard deviation; (b) outlier filtering; and (c) location determination (see Figure 2).

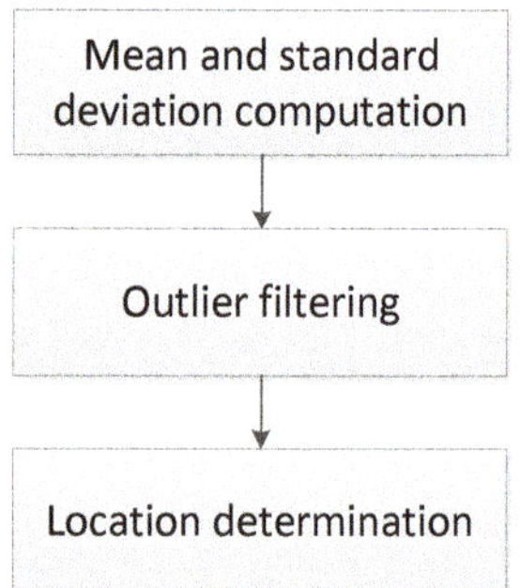

Figure 2. The steps of the mobile positioning method.

Mean and Standard Deviation Computation

For computation of the center, this study uses Equations (1) and (2) to calculate the means and standard deviations of longitudes and uses Equations (3) and (4) to calculate the means and standard deviations of latitudes, respectively:

$$\mu_{x,i} = \frac{\sum\limits_{j=1}^{r_i} x_{i,j}}{r_i} \tag{1}$$

$$\sigma_{x,i} = \frac{\sum\limits_{j=1}^{r_i} \left(x_{i,j} - \mu_{x,i}\right)^2}{r_i - 1} \tag{2}$$

$$\mu_{y,i} = \frac{\sum_{j=1}^{r_i} y_{i,j}}{r_i} \tag{3}$$

$$\sigma_{y,i} = \frac{\sum_{j=1}^{r_i} \left(y_{i,j} - \mu_{y,i}\right)^2}{r_i - 1} \tag{4}$$

Outlier Filtering

This step performs an outlier filtering mechanism to clean historical CVO data to determine location information precisely. The mechanism uses a threshold t that is defined as an outlier filtering criterion. The value of t can be set according to standard deviations $\sigma_{x,i}$ and $\sigma_{y,i}$. The longitude of record $x_{i,j}$ will be filtered out when it is smaller than $\mu_{x,i} - t$ or larger than $\mu_{x,i} + t$. Latitude records $y_{i,j}$ can be filtered with the same threshold t.

Location Determination

After outlier filtering, the means of the longitude and latitude of each record are calculated with Equations (5) and (6), respectively. This method determines the longitude and latitude of the i-th cell-RSSI pair as $\mu_{x,i}'$ and $\mu_{y,i}'$:

$$\mu_{x,i}' = \frac{\sum_{j=1}^{r_i'} x_{i,j}'}{r_i'}, \text{where } \mu_{x,i} - t < x_{i,j}' < \mu_{x,i} + t \tag{5}$$

$$\mu_{y,i}' = \frac{\sum_{j=1}^{r_i'} y_{i,j}'}{r_i'}, \text{where } \mu_{y,i} - t < y_{i,j}' < \mu_{y,i} + t \tag{6}$$

In these steps, the means and standard deviations can be calculated using cloud computing (e.g., the MapReduce programming model). The key is the cell-RSSI pair, and the values are longitudes and latitudes. The MapReduce programming model can be implemented for center computation.

3.1.3. Output Data

The location of each cell-RSSI pair is the output of the mobile positioning algorithm. This information can be calculated in the pre-deployment stage and stored in a cloud computing database for the runtime stage. The HBase technique [31] is used to implement this cloud computing database, and the HIVE technique [32] is used to perform the operations of cloud computing database.

3.2. Runtime Stage

In this stage, the ID of the connected cell and RSSI of the OBU can be retrieved by the CVO server when GPS is unavailable. The cell-RSSI pair obtained with the cell ID and RSSI is then used to query the pre-deployment cloud database for real-time location determination.

4. Analytical Models with Different Outlier Filtering Criteria

This section proposes an analytical model and presents numerical results to analyze the feasibility of the proposed mobile positioning method with different outlier filtering criteria.

4.1. Analytical Models

This section proposes models to analyze the relationships of location error and availability to different outlier filtering thresholds t.

4.1.1. Location Error

The locations of the same cell-RSSI pair are assumed to be normally distributed, and the expected ranges of the longitude and latitude of the i-th cell-RSSI pair are defined as $d_{x,i}$ and $d_{y,i}$ (see Equations (7) and (8)). The location errors may be generated in accordance with these ranges:

$$
\begin{aligned}
d_{x,i} &= \int_{\mu_{x,i}-t}^{\mu_{x,i}+t} \left| x_i - \mu_{x,i} \right| f\left(x_i, \mu_{x,i}, \sigma_{x,i}\right) dx_i, \text{ where } f\left(x_i, \mu_{x,i}, \sigma_{x,i}\right) = \frac{1}{\sigma_{x,i}\sqrt{2\pi}} e^{-\frac{(x_i-\mu_{x,i})^2}{2\sigma_{x,i}^2}} \\
&= \int_{\mu_{x,i}}^{\mu_{x,i}+t} \left(x_i - \mu_{x,i}\right) \left(\frac{1}{\sigma_{x,i}\sqrt{2\pi}} e^{-\frac{(x_i-\mu_{x,i})^2}{2\sigma_{x,i}^2}} \right) dx + \\
&\qquad \int_{\mu_{x,i}-t}^{\mu_{x,i}} \left(\mu_{x,i} - x_i\right) \left(\frac{1}{\sigma_{x,i}\sqrt{2\pi}} e^{-\frac{(x_i-\mu_{x,i})^2}{2\sigma_{x,i}^2}} \right) dx \\
&= \frac{2\sigma_{x,i}\left(1-e^{-\frac{t^2}{2\sigma_{x,i}^2}}\right)}{\sqrt{2\pi}}
\end{aligned}
\tag{7}
$$

$$
\begin{aligned}
d_{y,i} &= \int_{\mu_{y,i}-t}^{\mu_{y,i}+t} \left| y_i - \mu_{y,i} \right| f\left(y_i, \mu_{y,i}, \sigma_{y,i}\right) dy_i, \text{ where } f\left(y_i, \mu_{y,i}, \sigma_{y,i}\right) = \frac{1}{\sigma_{y,i}\sqrt{2\pi}} e^{-\frac{(y_i-\mu_{y,i})^2}{2\sigma_{y,i}^2}} \\
&= \int_{\mu_{y,i}}^{\mu_{y,i}+t} \left(y_i - \mu_{y,i}\right) \left(\frac{1}{\sigma_{y,i}\sqrt{2\pi}} e^{-\frac{(y_i-\mu_{y,i})^2}{2\sigma_{y,i}^2}} \right) dy + \\
&\qquad \int_{\mu_{y,i}-t}^{\mu_{y,i}} \left(\mu_{y,i} - y_i\right) \left(\frac{1}{\sigma_{y,i}\sqrt{2\pi}} e^{-\frac{(y_i-\mu_{y,i})^2}{2\sigma_{y,i}^2}} \right) dy \\
&= \frac{2\sigma_{y,i}\left(1-e^{-\frac{t^2}{2\sigma_{y,i}^2}}\right)}{\sqrt{2\pi}}
\end{aligned}
\tag{8}
$$

4.1.2. Availability

The locations of the same cell-RSSI pair are assumed to be normally distributed, and the availability of the i-th cell-RSSI pair is defined as $p_{x,i}$ and $p_{y,i}$ (see Equations (9) and (10)). The function $g(z)$ in Equations (9) and (10) is a Gaussian error function, which can be expressed as the Taylor series [37] $\frac{2}{\sqrt{\pi}} \sum_{n=0}^{\infty} \frac{z}{2n+1} \prod_{k=1}^{n} \frac{-z^2}{k}$ s [33]:

$$
\begin{aligned}
p_{x,i} &= \int_{\mu_{x,i}-t}^{\mu_{x,i}+t} f\left(x_i, \mu_{x,i}, \sigma_{x,i}\right) dx_i, \text{ where } f\left(x_i, \mu_{x,i}, \sigma_{x,i}\right) \\
&= \frac{1}{\sigma_{x,i}\sqrt{2\pi}} e^{-\frac{(x_i-\mu_{x,i})^2}{2\sigma_{x,i}^2}} \\
&= g\left(\frac{t}{\sqrt{2}\sigma_{x,i}}\right), \text{ where } g\left(z\right) = \frac{2}{\sqrt{\pi}} \sum_{n=0}^{\infty} \frac{z}{2n+1} \prod_{k=1}^{n} \frac{-z^2}{k} \\
&= \frac{2}{\sqrt{\pi}} \sum_{n=0}^{\infty} \frac{\frac{t}{\sqrt{2}\sigma_{x,i}}}{2n+1} \prod_{k=1}^{n} \frac{-\left(\frac{t}{\sqrt{2}\sigma_{x,i}}\right)^2}{k}
\end{aligned}
\tag{9}
$$

$$p_{y,i} = \int_{\mu_{y,i}-t}^{\mu_{y,i}+t} f\left(y_i, \mu_{y,i}, \sigma_{y,i}\right) dy_i, \text{ where } f\left(y_i, \mu_{y,i}, \sigma_{y,i}\right)$$

$$= \frac{1}{\sigma_{y,i}\sqrt{2\pi}} e^{-\frac{(y_i-\mu_{y,i})^2}{2\sigma_{y,i}^2}}$$

$$= g\left(\frac{t}{\sqrt{2}\sigma_{y,i}}\right), \text{ where } g(z) = \frac{2}{\sqrt{\pi}} \sum_{n=0}^{\infty} \frac{z}{2n+1} \prod_{k=1}^{n} \frac{-z^2}{k}$$

$$= \frac{2}{\sqrt{\pi}} \sum_{n=0}^{\infty} \frac{\frac{t}{\sqrt{2}\sigma_{y,i}}}{2n+1} \prod_{k=1}^{n} \frac{-\left(\frac{t}{\sqrt{2}\sigma_{y,i}}\right)^2}{k}$$

$$(10)$$

4.2. Numerical Analyses

To demonstrate the proposed analytical model, the following parameters were adopted to estimate the expected ranges of the longitude and latitude of the i-th cell-RSSI pair: $\mu_{x,i} = 120.3259728$, $\mu_{y,i} = 22.56716916$, $\sigma_{x,i} = 0.001271737$, $\sigma_{y,i} = 0.000940652$. Figure 3 shows the expected ranges of longitude with different outlier filtering thresholds t. The expected range is approximately 0.001003427 (*i.e.*, 103.03 m) when $t = 3 \times \sigma_{x,i}$ (*i.e.*, $t = 0.003815211$). Moreover, the expected range is $\frac{2\sigma_{x,i}}{\sqrt{2\pi}} = 0.001014699$ (*i.e.*, 104.19 m) when $t = \infty$. Therefore, the improvement of location determination is 1.16 m after outlier filtering with $t = 3 \times \sigma_{x,i}$. However, the availability $p_{x,i}$ decreases when the outlier filtering threshold t is decreased. Therefore, there is a trade-off between the location error and availability.

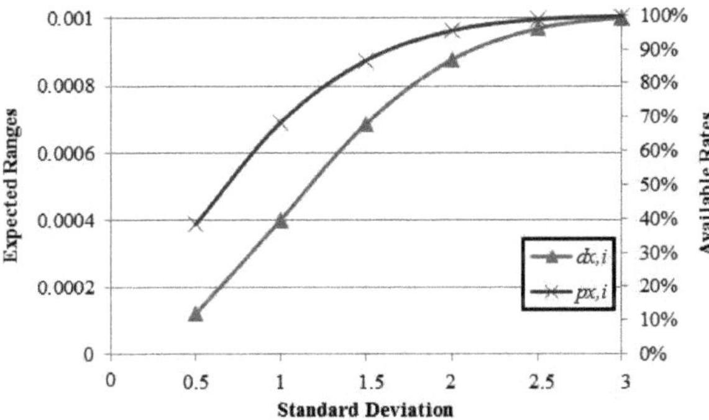

Figure 3. The expected ranges of longitude with different outlier filtering thresholds.

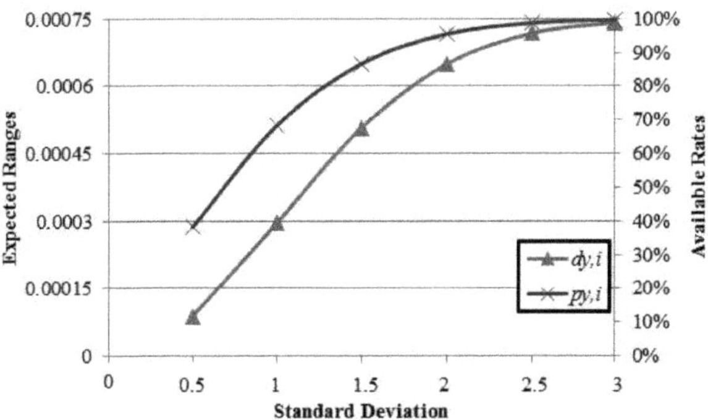

Figure 4. The expected ranges of latitude with different outlier filtering thresholds.

Figure 4 shows the expected ranges of latitude with different outlier filtering thresholds t. The expected range is approximately 0.000742194 (*i.e.*, 82.53 m) when $t = 3 \times \sigma_{y,i}$ (*i.e.*, $t = 0.002821957$), and $\frac{2\sigma_{y,i}}{\sqrt{2\pi}} = 0.000750532$ (*i.e.*, 83.46 m) when $t = \infty$. Therefore, latitude location determination improves to 0.97 m after outlier filtering with $t = 3 \times \sigma_{y,i}$.

5. Implementation and Evaluation of a Commercial Vehicle Operation System

In this section, the architecture of a CVO system is proposed and implemented, and the experimental practical results from the CVO system are compared with different location determination methods to evaluate the proposed mobile positioning method [33].

5.1. CVO System

This subsection proposes a CVO system composed of OBUs, a CVO server, and a cloud computing database server (shown in Figure 5).

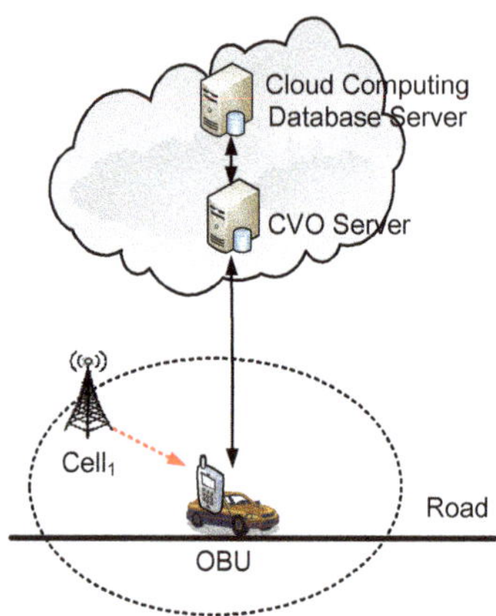

Figure 5. The architecture of the CVO system.

A GPS receiver and a cellular network module can be equipped in each OBU for periodically transmitting location information (*i.e.*, longitude and latitude) determined from GPS and cellular network signals (*i.e.*, the ID and RSSI of the connected cell) to the CVO server. When GPS is unavailable, the OBU sends only cellular network signals, which are presented as cell-RSSI pairs. The CVO server can perform the proposed mobile positioning method using the cell-RSSI pair as a key and querying the pre-deployment cloud computing database based on Hadoop [31], MapReduce [31], and Hive [32] techniques. The corresponding location of cell-RSSI pair can then be retrieved for determining the location of the OBU.

5.2. Experimental Results and Discussions

This subsection presents a case study and analyzes three months of CVO data to evaluate the proposed mobile positioning method. From November 2013 to January 2014, 67 OBUs were driven in experimental environments and 18,508 different cells were detected and connected. These OBUs obtained 6,571,550 CVO records and transmitted them to the CVO server for analyses of location information and cellular network signals.

5.2.1. Case Study

An OBU was selected on 23 November 2013 as a case study to present the results of the proposed mobile positioning method. The experiments were conducted on a highway segment 614 km long between Kaohsiung and Taoyuan in Taiwan (shown in Figure 6).

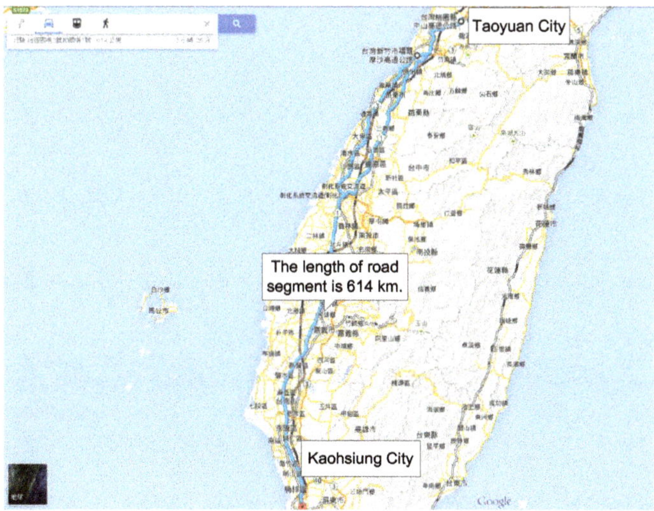

Figure 6. The experiment environment on Highways No. 1 and No. 3 in Taiwan on 23 November 2013.

Figures 7 and 8 show the location information of the OBU with different positioning methods. The green point locations were determined by GPS; the red points were determined by mobile positioning methods. In these cases, the GPS satellite signals were weak. Many locations determined with the cell-ID-based positioning method [9] are not properly on the road segment in Figure 7. The results show that the cell-ID-based positioning method cannot provide precise location information. However, Figure 8 shows that the locations determined using the proposed positioning method fit the road segment. Therefore, the proposed positioning method is more suitable than cell-ID-based positioning.

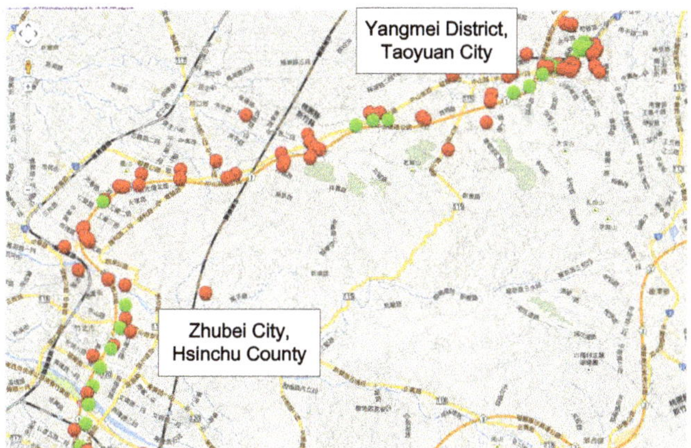

Figure 7. The results of the cell-ID-based positioning method on 23 November 2013.

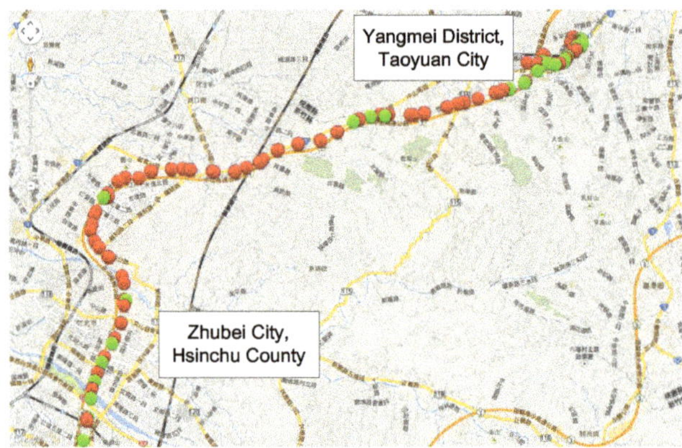

Figure 8. The results of the proposed positioning method on 23 November 2013.

5.2.2. Evaluation and Discussions

The practical results of using the proposed mobile positioning method are illustrated and evaluated in this section. In this study, the CVO data from October 2013 to January 2014 was collected for evaluation. The data of October 2013 was used as training data, and the data from November 2013 to January 2014 was used as testing data.

For the analyses of OBU traces, Table 1 shows the practical results of location determination with different positioning methods (*i.e.*, GPS, cell-ID-based method, and the proposed method) on 23 November 2013. The location information obtained from a GPS receiver was defined as the baseline data. In this case, the availability of GPS was only approximately 6.31%, so 93.69% of total records would lack GPS location information. This study compared location information using mobile positioning methods and GPS when GPS was available. The cell-ID-based positioning method [9] was considered to be implemented and evaluated, and the errors in location determination and availability were approximately 521.2 m and 99.51%, respectively. Finally, this study implemented and evaluated the proposed mobile positioning method, and the errors of location determination and availability were approximately 163.7 m and 99.58%, respectively. The CDFs of location errors in this case are shown in Figure 9. In another case, Table 2 and Figure 10 compare different positioning methods on 4 December 2013. The results indicate that the availabilities of the cell-ID-based method and the proposed method are 72.50% and 99.49%, respectively. These results show that the proposed mobile positioning method can provide the precise location information and is suitable for CVO.

Table 1. The comparison of different positioning methods on 23 November 2013.

	GPS	**Cell ID Based Method**	**The Proposed Method**
Average Location Error (m)	Baseline	521.2	163.7
Availability	6.31%	99.51%	99.58%

This study evaluates the proposed method for the six road types defined in [38] (national highway, provincial highway, urban road, county road, village road, and alley). From November 2013 to January 2014, 6,571,550 records were collected, and the ratio of records for each road type was calculated in Table 3. Because commercial vehicles are usually driven on national highways and urban roads, the ratios of national highways and urban roads are 36.470% and 42.639%, respectively. Tables 4 and 5 show the analyses of average location error and availability rate using various location determination methods. The practical experimental results indicate that the average location error using the proposed method is lower than using the cell-ID-based positioning method (Figure 11). The proposed method obtained more precise location information for each road type. Furthermore, when GPS is unavailable, the proposed method with its higher availability rate can be used to determine the locations of commercial vehicles for each road type, particularly urban roads.

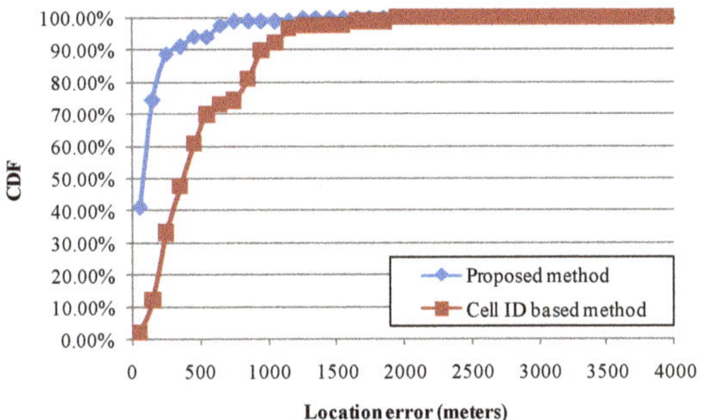

Figure 9. The comparison of different positioning methods on 23 November 2013.

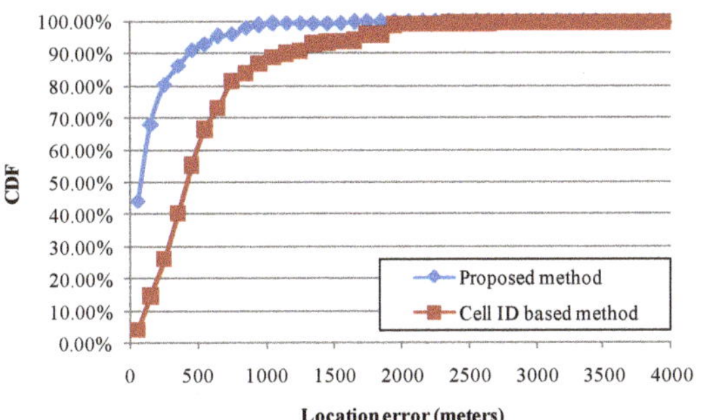

Figure 10. The comparison of different positioning methods on 4 December 2013.

Table 2. The comparison of different positioning methods on 4 December 2013.

	GPS	Cell ID Based Method	The Proposed Method
Average Location Error (m)	Baseline	597.8	193.4
Availability	99.26%	72.50%	99.49%

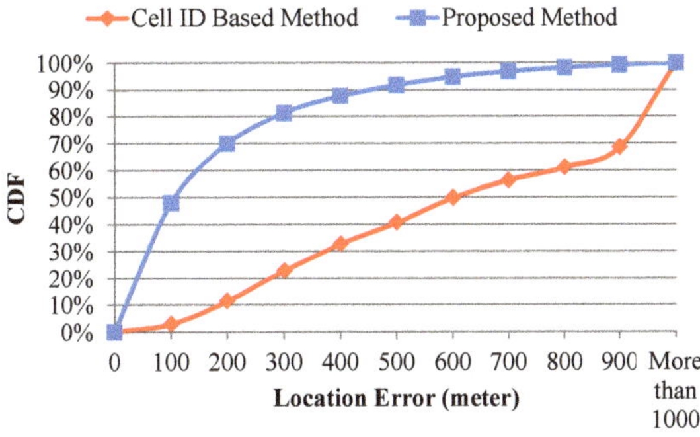

Figure 11. The comparison of different positioning methods.

Table 3. The ratio of historical records for each road type.

Road Type	The Number of Records	Ratio
National Highway	2,396,671	36.470%
Provincial Highway	559,433	8.513%
Urban Road	2,802,047	42.639%
County Road	275,894	4.198%
Village Road	237,131	3.608%
Alley	300,374	4.571%
Total	6,571,550	100%

Table 4. The analyses of average location errors for each road type (unit: m).

Road Type	Cell ID Based Method	The Proposed Method
National Highway	858.3	249.6
Provincial Highway	444.4	133.6
Urban Road	458.4	133.2
County Road	918.7	53.8
Village Road	324.6	49.3
Alley	474.8	132.9
Average	705.0	176.1

Table 5. The analyses of availabilities for each road type.

Road Type	GPS	Cell ID Based Method	The Proposed Method
National Highway	99.996%	99.998%	99.998%
Provincial Highway	99.997%	99.999%	99.999%
Urban Road	95.296%	99.951%	99.943%
County Road	99.990%	99.996%	99.996%
Village Road	99.994%	99.999%	99.999%
Alley	99.977%	99.994%	99.994%
Average	99.208%	99.989%	99.988%

6. Conclusions

A high-efficiency mobile positioning method is proposed to collect and analyze the cellular network signals of CVO data. Parallel computing and cloud computing techniques are designed into the proposed mobile positioning method to quickly determine the location of an OBU for CVO. Furthermore, this study proposes analytical models to analyze the availability of the proposed mobile positioning method with different outlier filtering criteria. In experimental environments, a CVO system was designed and implemented to collect CVO data from Chunghwa Telecom and to analyze the cellular network signals of CVO data for location determination. A case study determined that the average location errors using the proposed method and the traditional cell-ID-based method were 163.7 m and 521.2 m, respectively. Furthermore, the practical results show that the average location error and availability of using the proposed method are better than using GPS and the cell-ID-based method for each road type, particularly urban roads. Therefore, this approach is feasible to determine the location of an OBU for CVO improvement.

In future work, the signals of neighboring cells can be analyzed simultaneously to improve mobile positioning. Moreover, the proposed method for generating precise location information can be applied to support other ITS applications (e.g., advanced public transportation services, advanced traffic information services, *etc.*).

Acknowledgments: We would like to thank editor and reviewers for their comments. We also thank MDPI publisher for their supports.

Author Contributions: Chi-Hua Chen and Kuen-Rong Lo conceived and designed the experiments; Chi-Hua Chen performed the experiments; Chi-Hua Chen and Jia-Hong Lin analyzed the data; Ta-Sheng Kuan contributed analysis tools; Chi-Hua Chen wrote the paper.

Conflicts of Interest: The authors declare no conflict of interest.

References

1. Hsu, C.L.; Lin, J.C.C. A study of the adoption behavior for In-Car GPS navigation systems. *Int. J. Mob. Commun.* **2010**, *8*, 603–624. [CrossRef]
2. Liu, L.M.N.Y.; Lau, Y.C.; Patil, A.P. LANDMARC: Indoor location sensing using active RFID. *Wirel. Netw.* **2004**, *10*, 701–710.
3. Zhou, S.; Pollard, J.K. Position measurement using Bluetooth. *IEEE Trans. Consum. Electron.* **2006**, *52*, 555–558. [CrossRef]
4. Roos, T.; Myllymaki, P.; Tirri, H.; Misikangas, P.; Sievanan, J. A probabilistic approach to WLAN user location estimation. *Int. J. Wirel. Inf. Netw.* **2002**, *9*, 155–164. [CrossRef]
5. Youssef, M.; Agrawala, A. The Horus WLAN location determination system. In Proceedings of the 3rd International Conference on Mobile Systems, Applications, and Services, Seattle, Washington, DC, USA, 6–8 June 2005; pp. 205–218.
6. Chiou, Y.S.; Wang, C.L.; Yeh, S.C. An adaptive location estimator using tracking algorithms for indoor WLANs. *Wirel. Netw.* **2010**, *16*, 1987–2012. [CrossRef]
7. Guarnieri, A.; Pirotti, F.; Vettore, A. Low-cost MEMS sensors and vision system for motion and position estimation of a scooter. *Sensors* **2013**, *13*, 1510–1522. [CrossRef] [PubMed]
8. Cenedese, A.; Ortolan, G.; Bertinato, M. Low-density wireless sensor networks for localization and tracking in critical environments. *IEEE Trans. Veh. Technol.* **2010**, *59*, 2951–2962. [CrossRef]
9. 3GPP. *Technical Specification Group (TSG) Services and System Aspects*; TS 22.071; 3GPP: Valbonne, France, 2015.
10. Liu, H.; Darabi, H.; Banerjee, P.; Liu, J. Survey of wireless indoor positioning techniques and systems. *IEEE Trans. Syst. Man Cybern.* **2007**, *37*, 1067–1080. [CrossRef]
11. Open Mobile Alliance: Secure User Plane Location V2.0 Enabler Release Package. Available online: http://member.openmobilealliance.org/ftp/Public_documents/LOC/Permanent_documents/ OMA-ERP-SUPL-V2_0-20080627-C.zip (accessed on 26 March 2016).
12. Venkatachalam, M.; Etemad, K.; Ballantyne, W.; Chen, B. Location services in WiMAX networks. *IEEE Commun. Mag.* **2009**, *47*, 92–98. [CrossRef]
13. Cong, L.; Zhuang, W. Hybrid TDOA/AOA mobile user location for wideband CDMA cellular systems. *IEEE Trans. Wirel. Commun.* **2002**, *1*, 439–447. [CrossRef]
14. Qi, Y.; Kobayashi, H.; Suda, H. Analysis of wireless geolocation in a non-line-of-sight environment. *IEEE Trans. Wirel. Commun.* **2006**, *5*, 672–681.
15. Niculescu, D.; Nath, B. *Ad Hoc* Positioning System (APS) using AOA. In Proceedings of the IEEE INFOCOM Twenty-Second Annual Joint Conference of the IEEE Computer and Communications, San Francisco, CA, USA, 30 March–3 April 2003; pp. 1734–1743.
16. Addlesee, M.; Curwen, R.; Hodges, S.; Newman, J.; Steggles, P.; Ward, A.; Hopper, A. Implementing a sentient computing system. *Computer* **2001**, *34*, 50–56. [CrossRef]
17. Savvides, A.; Han, C.C.; Strivastava, M.B. Dynamic fine-grained localization in ad-hoc networks of sensors. In Proceedings of the ACM/IEEE MOBICOM International Conference on Mobile Computing and Networking, Rome, Italy, 16–21 July 2001; pp. 166–179.
18. Bshara, M.; Orguner, U.; Gustafsson, F.; Biesen, L.V. Fingerprinting localization in wireless networks based on received-signal-strength measurements: A case study on WiMAX networks. *IEEE Trans. Veh. Technol.* **2010**, *59*, 283–294. [CrossRef]
19. Bshara, M.; Orguner, U.; Gustafsson, F.; Biesen, L.V. Robust tracking in cellular networks using HMM filters and cell-ID measurements. *IEEE Trans. Veh. Technol.* **2011**, *60*, 1016–1024. [CrossRef]
20. Chang, M.F.; Chen, C.H.; Lin, Y.B.; Chia, C.Y. The frequency of CFVD speed report for highway traffic. *Wirel. Commun. Mob. Comput.* **2015**, *15*, 879–888. [CrossRef]
21. Gundlegård, D.; Karlsson, J.M. Handover location accuracy for travel time estimation in GSM and UMTS. *IET Intell. Transp. Syst.* **2009**, *3*, 87–94. [CrossRef]

22. Paek, J.; Kim, K.H.; Singh, J.P.; Govindan, R. Energy-efficient positioning for smartphones using Cell-ID sequence matching. In Proceedings of the 9th International Conference on Mobile Systems, Applications, and Services, Bethesda, MD, USA, 28 June–1 July 2011; pp. 293–306.
23. Chen, C.H.; Lin, B.Y.; Lin, C.H.; Liu, Y.S.; Lo, C.C. A green positioning algorithm for campus guidance system. *Int. J. Mob. Commun.* **2012**, *10*, 119–131. [CrossRef]
24. Lin, B.Y.; Chen, C.H.; Lo, C.C. A novel speed estimation method using location service events based on fingerprint positioning. *Adv. Sci. Lett.* **2011**, *4*, 3735–3739. [CrossRef]
25. Chen, C.H.; Lo, C.C.; Lin, H.F. The analysis of speed-reporting rates from a cellular network based on a fingerprint-positioning algorithm. *S. Afr. J. Ind. Eng.* **2013**, *24*, 98–106. [CrossRef]
26. Chen, C.H.; Lin, B.Y.; Chang, H.C.; Lo, C.C. The novel positioning algorithm based on cloud computing—A case study of intelligent transportation systems. *Information* **2012**, *15*, 4519–4524.
27. Cheng, D.Y.; Chen, C.H.; Hsiang, C.H.; Lo, C.C.; Lin, H.F.; Lin, B.Y. The optimal sampling period of a fingerprint positioning algorithm for vehicle speed estimation. *Math. Prob. Eng.* **2013**, *2013*, 1–12. [CrossRef]
28. Wigren, T. Adaptive enhanced cell ID fingerprinting localization by clustering of precise position measurements. *IEEE Trans. Veh. Technol.* **2007**, *56*, 3199–3209. [CrossRef]
29. Kuo, S.P.; Lin, S.C.; Wu, B.J.; Tseng, Y.C.; Shen, C.C. GeoAds: A middleware architecture for music service with location-aware advertisement. In Proceedings of the IEEE International Conference on Mobile Ad-hoc and Sensor Systems, Pisa, Italy, 8–11 October 2007.
30. Chen, C.H. Traffic Information Estimation Methods Based on Cellular Network Data. Ph.D. Thesis, Department of Information Management and Finance, National Chiao Tung University, Hsinchu, Taiwan, 2013.
31. Apache Software Fundation, Apache Hadoop 2.3.0. 2015. Available online: http://hadoop.apache.org/ (accessed on 26 March 2016).
32. Apache Software Fundation, Apache Hive 0.12.0. 2015. Available online: http://hive.apache.org/ (accessed on 26 March 2016).
33. Chen, C.H.; Lin, J.H.; Kuan, T.S.; Lo, K.R. A high-efficiency mobile positioning system by using commercial vehicle operation data based on cloud computing techniques. In Proceedings of the IEEE International Conference on Internet of Things, Taipei, Taiwan, 1–3 September 2014.
34. Chen, C.H.; Lin, S.Y.; Chang, H.C.; Lo, C.C. On the design and development of a novel real-time transaction price estimation system. *Adv. Mater. Res.* **2011**, *393–395*, 213–216. [CrossRef]
35. Levine, D.; Krehbiel, T.C.; Berenson, M.L. *Basic Business Statistics: Concepts and Applications*, 10th ed.; Pearson Education: New York, NY, USA, 2005.
36. Chen, C.H.; Lin, H.F.; Chang, H.C.; Ho, P.H.; Lo, C.C. An analytical framework of a deployment strategy for cloud computing services: A case study of academic websites. *Math. Prob. Eng.* **2013**, *2013*, 1–14. [CrossRef]
37. Wikipedia, Error Function. 2015. Available online: http://en.wikipedia.org/wiki/Error_function (accessed on 26 March 2016).
38. Ministry of Justice of the Republic of China, Highway Act. 2015. Available online: http://law.moj.gov.tw/ LawClass/LawAll.aspx?PCode=K0040001 (accessed on 26 March 2016).

International Journal of
Geo-Information

Article

Efficient Location Privacy-Preserving *k*-Anonymity Method Based on the Credible Chain

Hui Wang [1,2], Haiping Huang [1,2,3,4,*], Yuxiang Qin [1,2], Yunqi Wang [1,2] and Min Wu [1,2,3]

1   School of Computer Science, Nanjing University of Posts and Telecommunications, Nanjing 210003, China;
    hughwangmail@yeah.net (H.W.); 18705192253@163.com (Y.Q.);
    wangyunqi773@163.com (Y.W.); wumin@njupt.edu.cn (M.W.)
2   Jiangsu High Technology Research Key Laboratory for Wireless Sensor Networks, Nanjing 210003, China
3   College of Computer Science and Technology, Nanjing University of Aeronautics and Astronautics,
    Nanjing 210016, China
4   Institute of Computer Technology, Nanjing University of Posts and Telecommunications,
    Nanjing 210003, China
*   Correspondence: hhp@njupt.edu.cn

Academic Editors: Chi-Hua Chen, Kuen-Rong Lo and Wolfgang Kainz
Received: 9 December 2016; Accepted: 30 May 2017; Published: 1 June 2017

Abstract: Currently, although prevalent location privacy methods based on *k*-anonymizing spatial regions (K-ASRs) can achieve privacy protection by sacrificing the quality of service (QoS), users cannot obtain accurate query results. To address this problem, it proposes a new location privacy-preserving *k*-anonymity method based on the credible chain with two major features. First, the optimal *k* value for the current user is determined according to the user's environment and social attributes. Second, rather than forming an anonymizing spatial region (ASR), the trusted third party (TTP) generates a fake trajectory that contains *k* location nodes based on properties of the credible chain. In addition, location-based services (LBS) queries are conducted based on the trajectory, and privacy level is evaluated by instancing θ privacy. Simulation results and experimental analysis demonstrate the effectiveness and availability of the proposed method. Compared with methods based on ASR, the proposed method guarantees 100% QoS.

Keywords: *k*-anonymity; location-based services; location privacy; the credible chain

---

1. Introduction

As one of the most important forms of digital information, geographical location data play a critical role in various applications (e.g., smart cities, social networks and intelligent navigation) via big data processing, mobile communications and sensing technologies. Consequently, location-based services (LBS) have become some of the most prevalent tools used in all kinds of Internet of things' applications. Many location applications can be downloaded via the applications market through users' smart phones or tablet computers. With the help of these applications, users can easily obtain location query services and relevant points of interest (POI) returned by a location server. For example, users can query nearby hospitals, restaurants or gas stations.

Location data can disclose private personal information while offering convenience to users; as such, data not only include user location coordinates but also reveal other sensitive personal data such as users' habits, health conditions, and social affiliations [1]. The abuse of location information can considerably compromise user privacy. Several ways to address such issues of location privacy have been proposed over the past few years. Such methods can be divided into two categories: those based on the location privacy-preserving model with the trusted third party (TTP) and those based on the location privacy-preserving model without TTP. The privacy-preserving model without TTP exacerbates communication costs, delays and computational complexity levels and presents obvious problems of usability and stability. The location privacy-preserving model based on TTP is consequently more suitable in use in practical scenarios in combination with a trusted third party service [2]. The model adds a firewall between the user and LBS server and uses location perturbation and obfuscation to achieve privacy protection, and most commonly via *k-anonymity*. To achieve *k-anonymity*, TTP expands the queried location into a broader anonymizing spatial region (ASR) that covers several other users (e.g., other $k - 1$ users) geographically. As a result, it is difficult for an untrusted LBS server to determine a user's real location from other $k - 1$ dummy locations [3]. However, these approaches based on *k*-anonymity achieve high-level privacy protection while sacrificing service quality levels. While a broader ASR achieves greater user privacy protection, it occurs at the cost of lower service quality and higher communication and computation costs. Therefore, the trade-off between privacy protection, service quality and resource costs is a major concern with respect to location privacy protection technologies.

This paper proposes a location privacy-preserving method of *k*-anonymity based on the credible chain. The method should not affect the QoS of LBS, as it adopts a trajectory that protects location privacy rather than constructing an ASR. This paper adopts a classical LBS system architecture based on a central anonymity server (anonymizer) and determines the best *k* value for a user based on the user's environment and social attributes. To address contradictions between privacy protection levels, service quality and resource costs, it utilizes properties of the credible chain to predict the next state and constructs an illusive location trajectory that contains *k* locations.

The main contributions of this paper can be summarized as follows:

(1) It proposes a new location privacy-preserving method of *k*-anonymity based on the credible chain. Rather than forming a *k*-ASR similar to existing schemes, a TTP forms a fake trajectory that includes *k*-locations based on properties of the credible chain. It can achieve 100% service accuracy while protecting user location privacy.

(2) A feasible *k* value selection scheme is proposed as a way to reduce unnecessary communication overhead while guaranteeing user location privacy. The *k* value is not static and is calculated in terms of a user's current environment and social attributes.

(3) Finally, it conducts privacy metrics by instancing θ privacy to validate the effectiveness and accessibility of this method. Compared with existing schemes, the proposed method achieves superior service performance.

The remainder of the paper is organized as follows. Related works are summarized in Section 2. The system model is introduced in Section 3. Section 4 describes the location privacy-preserving method of *k*-anonymity based on the credible chain in detail. Section 5 presents privacy metrics that involve instancing θ privacy. Section 6 presents the experimental analysis and performance evaluation. Finally, Section 7 draws conclusions.

2. Related Works

Several ways to ensure location privacy have been proposed [3–17].

As noted above, these approaches can be divided into two categories according to system structures: those related to the location privacy-preserving model based on TTP, which can protect a user's personal information via concealment or confusion, and those based on the location privacy-preserving model without TTP, which can be divided further into collaborative and non-collaborative methods.

For the former, many solutions have been proposed such as methods based on anonymous boxes and data features, for which *k*-anonymity is now the most widely used tool. Sweeney et al. [3] developed a *k*-anonymity model as a privacy protection measure for ensuring personal data privacy. In 2003, *k*-anonymity was first applied to location privacy protection by Gruteser et al. [4]. In [4], location perturbation of *k*-anonymity method is performed via the quadtree-based algorithm, which adopts spatial and temporal cloaking. However, this approach presents two drawbacks: (1) first, it uses a static k value as a privacy parameter for all mobile users, which likely affects service quality levels for those users whose privacy needs can likely be satisfied using a smaller k value. Furthermore, this assumption is unrealistic, as mobile users tend to have alterable privacy protection needs under different conditions and on different subjects. (2) Second, it can easily generate an excess anonymous region, which not only increases computation costs but also affects the quality of services. To address static k value issues, Gedik et al. [5] proposed the CliqueCloak method, which adopts an individualization k-anonymity model to protect location privacy. However, it only supports small k values (5–10) given its high degree of computational complexity. In [6], Bottom-Up Grid Cloaking and Top-Down Grid Cloaking methods are proposed as ways to form anonymous regions by respectively merging or decomposing grid regions. Bottom-Up Grid Cloaking is used to manage location queries with fewer privacy requirements and Top-Down Grid Cloaking is used to manage location queries with greater privacy requirements. Jagwani et al. [7] proposed a k-anonymity method based on fuzzy spatiotemporal contexts. While this method involves determining the k value to prevent location disclosure and using current fuzzy spatiotemporal attributes to guarantee a more reasonable k value, it does not account for a user's social attributes such as the correlation degree between identity and location and the associated number for others. To address excess anonymous regions issues, the Casper algorithm [8] was proposed by Mokbel as a way to form an ASR based on [4]. When a user's number of current quadtree leaf nodes where the request sender is located is less than k, the area of the current leaf node is merged with that of its adjacent sibling node. When a user's number of anonymous areas is still less than k, the area of the parent of the existing leaf node must be searched for. The Casper algorithm is superior to the algorithm shown in [4], as it reduces computation costs and allows users to more easily control privacy parameters. However, it still suffers from excess anonymous regions and unsatisfactory service quality levels. Yong et al. [9] present a location privacy-preserving k-anonymous method based on service similarities. The location service similarity is introduced to assist anonymity servers in looking for anonymous areas, which not only improves an individual's need for high-quality information services to some extent (however, not 100%), but also reduces the computation and communication overhead. Niu et al. [10] propose the Dummy-Location Selection (DLS) and enhanced-DLS algorithms. The DLS algorithm carefully selects dummy locations based on the entropy metric, as side information can be exploited by adversaries. The enhanced-DLS algorithm ensures that selected dummy locations are spread out as much as possible, and it can expand the cloaking region while maintaining privacy levels similar to those of the DLS algorithm.

Location privacy-preserving model without TTP consists of an LBS server and several mobile users, and mobile users form a fake location or k-ASR through a cooperative or independent way to meet the anonymous area requirement to achieve location privacy protection. Chow et al. [11] proposed a peer-to-peer spatial cloaking algorithm for anonymous location-based services. The main premise of this algorithm is that before requesting any location-based service, a mobile user must form a group based on his/her neighbor users. A user of the group is then selected to send a service request to the LBS server. The algorithm proposed in [11] has since been improved by Chow [12], who increased the system's availability by allowing a user to use his/her historical neighbor data and corresponding anonymity levels to achieve location privacy when enough time is available. Zakhary et al. [13] proposed an HSLPO (Social-aware Location-Privacy in Opportunistic mobile social networks) algorithm that can identify a users' social network and use it to obfuscate service requests and to hide the original sender's location. The HSLPO algorithm can achieve higher levels of location privacy and service quality than other algorithms in terms of success ratios. However, these methods cannot address inherent defects of the location privacy-preserving model without TTP (i.e., due to high communication or computation costs).

In addition, some new methods combined with anonymous chains were proposed [14–17]. Historical trajectories are used to form anonymous location chains in [14,15] in order to achieve privacy protection. Many existing chain structures have also been introduced into privacy protection, Markov chain is one of the examples. Kang et al. achieved the user's ID secure authentication with

Markov chain [16]. Montazeri et al. proved that Markov chain, as the result of stochastic process, can be used to simulate the users' location trajectory, and, meanwhile, it achieves perfect location privacy [17]. However, how to combine the Markov chain with the real world has not been explained in [17]. Different from these methods, this paper introduces the users' states and social attributes into an anonymization model and form a fake trajectory in real areas.

Most of the existing methods are devoted to protecting the location privacy with ASR, and some others (e.g., the proposal in this paper) try to achieve the win-win situation between the quality of service and the privacy protection. In this paper, it protects the privacy with a group of fake points and makes them up into a fake trajectory. With the help of anonymity algorithms, these fake points are creditable and not easily distinguishable from real points by the attackers. With this method, it can hide the user's true location information and keep the 100% quality of service at the same time.

3. Systems Model

When building a safe LBS model, three factors must be considered:

(1) The quality of service (QoS).
(2) When information (part of the database) disclosure occurs and LBS data are leaked, leaked information can be controlled as little as possible.
(3) When location-based services are taken over by an attacker, the attacker can be misled with false data.
(4) The LBS will adopt the timestamp from the received request messages to provide location services.

When the LBS is contacted without a TTP and information is sent directly, the accuracy of a given service can be ensured. However, this method is less secure when subjected to attacks. When the k-anonymous method is used, the real user's location is merely a point in an ASR, and the redundant area inevitably results in inaccuracy and QoS degradation. However, when an attacker takes over an entire LBS, he can likely navigate the ambiguous areas of real location by using a fuzzy user trajectory.

It is true that there are still some disadvantages in the privacy-preserving model with TTP. For example, all of the protections are in vain if the TTP is thoroughly taken over by the attacker. However, taking computation cost and convenience of network management into account, it is believed that the privacy-preserving model with TTP is still the better choice in most real situations compared with that without TTP whose computation completely relies on individuals' devices. Consequently, an assumption needs to be considered in this paper: TTP is trusted and secure. The purpose of this paper is mainly to protect the users' information from the potential security threats in LBS.

One strong solution involves combining a real location with fake locations via a TTP and sending this location data to an LBS. Once a reply from the LBS is received, the TTP can return the correct answer to the user. This can be carried out to ensure QoS and privacy at the same time. However, an experienced attacker can exclude fake points via logical plausibility analysis. For example, one cannot travel around a city in 10 min or stand in the middle of a lake. After applying such exclusions, user location data are probably exposed to the attacker. Therefore, the key to the success of this method involves generating 'trusted points' and forming a 'trusted trajectory'.

To apply this idea, it generates 'trusted points' and transforms a user's location into a fake trajectory rather than a dummy region. As is shown in Figure 1, a *true* node is a user's true location node, and the *fake*1, *fake*2, *fake*3, ... *fake*(k − 1) nodes are selected from a cloud server to form a fake trajectory. A cloud server is a server from a TTP that stores previous requests or realistic points and that can ensure that fake points are located in viable areas rather than in locations that a request cannot cover. A trajectory is a sequence of moving object location data sorted by time. Hence, the anonymity server must change the timing of nodes to allow the trajectory to confuse and distract attackers. In Figure 1, the *fake*1 node starts the trajectory followed by the *fake*2 node via the *true* node and finally the *fake*(k − 1) node. The resulting user position accuracy is superior to that of an ASR, and a user can achieve higher QoS while ensuring his/her location privacy.

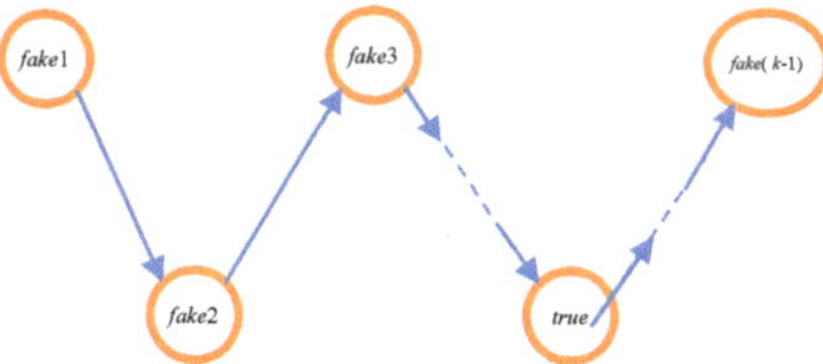

Figure 1. Diagrammatic sketch of the trajectory.

Based on Figure 1, it adopts the system structure shown in Figure 2. When a user sends a service request, the client determines the *k* value that meets the user's anonymous needs according to the user's environment and social attributes, and then the client sends this *k* value to the anonymity server. The anonymity server then obtains $k - 1$ fake nodes by communicating with the cloud server, and $k - 1$ fake nodes and the true node are constructed into a fake trajectory. Finally, the LBS server carries out inquiry processing for nodes of the fake trajectory in order.

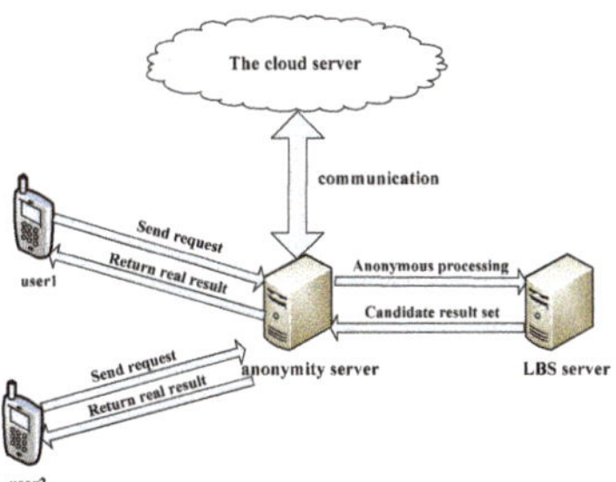

Figure 2. System structure.

4. *k*-Anonymity Method Based on the Credible Chain

Based on the above system model, it proposes a *k*-anonymous location privacy protection method based on the credible chain.

4.1. Preliminary Knowledge

Definition 1. *(Request message Q) the requested message Q can be expressed as a six tuple:*

$$Q = \{id, loc, t, qry, k, s\},$$

where id is the identity information of the user who sends the request; loc is the user's location, which can be directly obtained from a Global Position System (GPS) or using other positioning devices; t denotes the time at which the user sends the request; qry is the content that the user wants to request; k is the anonymous parameter of the location privacy level, which can be determined from the system; and s is the anonymous region where the user located. For example, if the user is in Shanghai, the history points, which will be chosen in following sections, can only be in the same city.

Definition 2. *(The credible chain) Let $\{X(t), t \geq 0\}$ denote a discrete time process taking values in state space I $= \{0,1,2,\dots\}$. For $0 \leq t_1 < t_2 < \dots < t_{n+1}$ and $i_1, i_2, \dots, i_{n+1} \in I$, $P\{X(t_{n+1}) = i_{n+1} \mid X(t_1) = i_1, X(t_2) = i_2, \dots, X(t_n) = i_n\} = P\{X(t_{n+1}) = i_{n+1} \mid X(t_n) = i_n\}$. Note that $\{X(t), t \geq 0\}$ is defined as the credible chain.*

When the locations where the user is positioned at the present time t_n and at all past times are known, the location where the user is located in the future t_{n+1} is only related to t_n. In addition, $P\{X(t_{n+1}) = i_{n+1} \mid X(t_n)$

$= i_n\} \neq 0$ denotes that from the location where the user is currently located at time t_n, the user can arrive at the location where he/she will be positioned at time t_{n+1}.

There are two advantages to adopting the credible chain:

(1) The unit of a credible chain is only related to the units preceding it. This ensures the uniformity of the entire trajectory and renders it more 'trusted'.

(2) Due to the non-aftereffect properties of the credible chain, it is impossible for an attacker to identify previous points according to leaked data.

Definition 3. *(Trajectory based on the credible chain T) The trajectory is generated by a TTP and includes at least k request messages (containing a user request message), the location of q_i and the next reachable location of q_{i+1} (i = 0,1, … ,k − 1) satisfy the inequality $P\{X(t_{i+1}) = q_{i+1}\ .loc\ |\ X(t_i) = q_i\ .loc\ \} \neq 0$.*

Definition 4. *(Query function R) R (loc) is a function that queries POI according to loc, and it can obtain a sorted set of top-m POIs. Sorting rules can be customized by an LBS (e.g., distance, popularity, rate and quality of service). The Euclidean Distance between the user and POI is adopted as the sorting rule.*

Definition 5. *(Candidate result set W) W denotes the set of all query results searched by the LBS based on location nodes of the trajectory provided by the anonymity server.*

4.2. k Value Selection

k is an important value in this model that represents the anonymous parameter of the location privacy level and that can be calculated through the system. It also denotes the number of points in the fake trajectory.

To further improve location privacy outcomes, the optimal user k value must be determined. There are a lot of factors that may influence the disclosure of the users' privacy information. Sometimes, the users' personal requirements need to be considered. According to users' location privacy requirements, investigation and analysis [18], in this paper, the four most prevalent factors are chosen which can be directly analyzed with the TTP's database or set by the users without the help of other data sources or technical tools. It is worth stressing that these factors are not essential to all anonymous scenarios, which can be substituted or added in terms of actual demands and situations, and they should not affect the effectiveness and availability of the algorithm.

(1) Density of the anonymous area

The density of the anonymous area (crowdedness) has a strong effect on location privacy. While individuals do not wish to expose themselves to less crowded areas, they may feel relatively safe in crowded areas. The less the density of the anonymous area is, the more important the location information is. In this paper, anonymous area density levels can be classified into four categories: sparse, moderately crowded, crowded, and extremely crowded. Such classifications are not fixed and can be altered according to realistic conditions. For example, the number of levels can be three, five, or greater. Level classifications do not affect the validity of the algorithm.

(2) Time interval of one day

Different users have different location privacy needs for different time intervals based on distinguished social attributes. For example, at night, individuals who work during the day usually have greater location privacy needs than those who work at night, and they thus require larger k values. Time intervals for a single day can be classified into 4 levels: night, morning, afternoon and evening.

(3) Correlation between identity and location

Users often have different location privacy needs even when they are located in the same area because, in certain environments, their locations are closely related to their identities. When relationships between user identity and location are stronger, smaller k values are required. For example, when a teacher queries an LBS on a campus, the value of k should be smaller than that used when the teacher is located in a bar. For example, correlation levels can be classified into four levels: irrelevant, low, moderate, and high.

(4) Associations with others

In forming social networks, many users determine their respective social circles. Associations with others denote the extent of a user's social circle. The larger the number of associations becomes, the stronger privacy-preserving demands become. For example, movie stars always keep their locations private, as their fans would bloat their number of associations, thus requiring a larger k value. Here, the number of associations can be divided into four levels: few, some, many and numerous.

The above four factors are considered to strongly influence user location privacy. Linguistic variables are used to describe the influence levels of the four factors. The linguistic variable is typically a fuzzy value (e.g., "few" or "many") rather than an accurate value (0, 1, 2, 3, etc.), thus allowing users to make decisions more intuitively. The number of levels for factors can be adjusted according to the user experience and actual demands of applications. The users have chances to choose their own levels of refinement. Actually, it does not affect the validity of the algorithm. These factors cannot be directly applied due to their varying scales. To eliminate the influence from the scales of different factors, the weight of each factor must be scientifically measured. The value of k will be calculated in turn based on these weights.

When calculating weights, it is necessary to obtain a large number of surveyed samples regarding the factors on a user's location privacy. In order to reflect different users' social attributes, these factors are quantified into specific values X_{ij} (the value of the ith factor of the jth sample). In this case, X_{ij} can be 0, 1, 2, 3, ... , $n-1$, which denote the number of levels of each factor. Then, average the value of each factor via Equation (1):

$$\overline{X}_i = \frac{\sum_{j=1}^{N} X_{ij}}{N} (i = 1, 2, 3, 4) \tag{1}$$

where N represents the total amount of surveyed samples selected from the database.

The weight of each factor can be standardized via Equation (2):

$$W_i = \frac{\overline{X}_i}{\overline{X}_1 + \overline{X}_2 + \overline{X}_3 + \overline{X}_4} (i = 1, 2, 3, 4) \tag{2}$$

Thus, the weights can be between 0 and 1.

It can calculate a synthetic factor $\sum_{i=1}^{4}(W_i * U_i)$ whose value lies in the range $[0, n-1]$ due to different emphases from four factors, where U_i denotes the value of the ith factor of the current user. Based on the synthetic factor value, it is important to find the most optimal k value for each user between K_{max} and K_{min}. The anonymous value k (i.e., the number of location nodes in the credible chain) is calculated according to Equation (3):

$$k = \left\lceil \frac{k_{max} - k_{min}}{n} * \left(\sum_{i=1}^{4}(W_i * U_i) - 1 \right) \right\rceil + k_{min}. \tag{3}$$

The lower and upper bound values of anonymous levels are expressed as k_{max} and k_{min}, respectively, and both of them can be set according to the specific situation by the anonymity server. The results in Equation (3) should round upwards to the nearest integer because k is an integer.

In Equation (3), the value of k is directly proportional to the values of the user's attributes. The more sensitive user data is the higher privacy level user needs (the value of k is bigger). However, the specific relationship between k and the factor values is various and alternative. A questionnaire has been conducted to investigate the user's privacy requirements and the corresponding results have been taken into linear regression to prove the feasibility to use linear dependency.

4.3. Anonymous Processing

Anonymous processing is a critical step of the location privacy-preserving model, where a user's location is involved in a fake trajectory based on the credible chain generated via an anonymity server. In this paper, two anonymous parameters (k and s) can be set according to a user's demands and background knowledge to satisfy his/her personalized location privacy-preserving needs. The procedure described above can be explained via Algorithm 1.

Algorithm 1 Make a fake trajectory

Input: The user's request q
Output: array P
1:    $q_{1:k-1} \leftarrow k - 1$ messages selected by anonymity server
2:    $p_{1:k} \leftarrow$ RANDOMSHUFFLE $(q, q_1, \ldots, q_{k-1})$ // The details of this step are shown in Algorithm 2
3:    // The following steps are just outlines, the details of the following part are shown in Algorithm 3
4:    **for** $i = 2, 3 \ldots k$ **do**
5:       $T \leftarrow$ NECESSARYTRAVELTIME (p_{i-1}, p_i)
6:       **If** $T \leq t_{p_i} - t_{p_{i-1}}$
7:       $t_{p_i} \leftarrow t_{p_i} + T + RandomDelay$
8:       **End If**
9:    **End for**
10:    **Return** P

The following two algorithms will show the details of some parts of Algorithm 1.

In Algorithm 2, $k - 1$ messages will be selected from the server. Then, the true request message will be mixed into them according to the request time.

Algorithm 2 Initial anonymous processing algorithm

Input: The anonymity server selects $k - 1$ messages from the cloud server according to s generated by the user and then places $k - 1$ messages and the user's message q_0 into the array Q (namely Q = $\{q_0, q_1, q_2, \ldots, q_{k-1}\}$).
Output: array P
11:    $date = q_0.t$ // Assign $q_0.t$ (the time of the user's request message) to the variable *date*.
12:    $i = 1$
13:    $\Delta T = \max (q_0 - q_i), i \in [1, k - 1]$
14:    **while** $i \leq k - 1$ **do** // Update the $k - 1$ message selected from the cloud server in turn.
15:       $q_i.id = q_0.id$
16:       $q_i.k = q_0.k$
17:       $q_i.s = q_0.s$
18:       **if** $qi.t \notin (date - \text{random}(\Delta T), date)$ **then**
19:        $qi.t = date + \text{random} (\Delta T)$
20:       **end if** // Substitute $date + \text{random}(\Delta T)$ for $qi.t$ while $qi.t$ does not fall within the range of $(date - \text{random}(\Delta T), date)$. The function of random(ΔT) is used to generate a random number in the range of $(0, \Delta T)$, and the random number is retained to one decimal place.
21:       $i = i + 1$
22:    **end while**
23:    $P = \text{Sort} (\{Q - q_i\})$ // Place these messages from the array Q into the other array P after sorting according to the value of t.
24:    **Return** P

In the process of selection, the region s can be divided into k sub-regions with the same size. For each sub-region where the current user is not located, it picks up the most inactive (i.e., the least frequently used recently) historical request message from the sub-region. If there is no inactive message in this sub-region, then one from the nearest sub-region can be borrowed. Thus, it can finally select $k - 1$ fake messages as the input of Algorithm 2.

In this algorithm, messages that have been requested before $date -$ random (ΔT) will be put after the true request message and their request time will be reset as $date +$ random (ΔT). By this means, the true point can be mixed with those fake points. However, the request time of true point is the current time, and the attacker can recognize it easily. The fake points may not be accessible at the current speed within the current time interval, and they can be easily excluded by the attacker. Algorithm 3 is used to solve all of these problems.

Each point in this trajectory has a region that it can access with its current speed in the current time interval. The transition probability for each point in this accessible region is equal, while the transition probability out of this region is zero. In this paper, all of the fake points are historical points chosen from the cloud service. The transition probability of each point in the trajectory is $1/M$, M is the number of the historical points within the current point's accessible region (a round area with the radius of $S_{i,i+1}/v_{i,i+1}$). Any point of this trajectory is inaccessible if its last point's transition probability is 0. In this case, it needs to delay the request time of the current point to expand its last point's accessible region and make the current point accessible. The detailed steps are described as the following.

Algorithm 3 The credible chain algorithm

Input: array P = $\{p_0, p_1, p_2, \ldots, p_{k-1}\}$, *date*
Output: the user's trajectory T based on the credible chain
1:   T = $\{p_0\}$ // Initialization should be completed before the credible chain is formed, and p_0 is placed into the trajectory T.
2: **if** $p_0.t = date$
3:   *flag* = 0 // When the message p_0 is exactly the user's message, *flag* is set to 0. This denotes that the user's message has been added to the credible chain.
4: **else** *flag* = 1
5: **end if**
6: $\Delta = 0; i = 0$ // Δ is the interval between the time of the user's request message after anonymity and the real time of the user's request message.
7: **while** $i \leq k - 2$ **do**
8:    **if** $\Delta t = S_{i,i+1}/v_{i,i+1} > (p_{i+1}.t - p_i.t)$ **do** // $p_i.loc$ cannot arrive at $p_{i+1}.loc$.
9:     **if** $p_{i+1}.t - \Delta = date$ && *flag* **then** // Judge whether p_{i+1} is the user's message.
10:      $\Delta = p_i.t + \Delta t - date$
11:     **end if**
12:     $p_{i+1}.t = p_i.t + \Delta t$ // $p_{i+1}.t$ is updated to guarantee that $p_i.loc$ can arrive at $p_{i+1}.loc$.
13:    **end if**
14:    add (p_{i+1}) // $p_i.loc$ can arrive at $p_{i+1}.loc$, illustrating that $P\{X(t_{i+1}) = q_{i+1}.loc \mid X(t_i) = q_i .loc\} \neq 0$. Hence, p_{i+1} should be added to the trajectory T.
15:    delete (p_{i+1}) // p_{i+1} should be removed from the array P.
16:    **when** $p_{i+1}.t - \Delta = date$, **then** // Determine whether p_{i+1} is the user's message.
17:     *flag* = 0; // if the user's message has been added to the credible chain, and *flag* should be set to 0.
18:    **end if**
19:   $i = i + 1$
20: **end while**
21: **Return** T

Algorithm 3 is executed after Algorithm 2, and the input for Algorithm 3 is obtained from the result of Algorithm 2. Initialization needs be carried out before the credible chain is formed. The critical step involves determining whether $\Delta t = S_{i,i+1}/v_{i,i+1} > (p_{i+1}.t - p_i.t)$ (where $S_{i,i+1}$ is the distance between $q_i.loc$ and $q_{i+1}.loc$ and $v_{i,i+1}$ is the maximum average speed at which the user arrives at $q_{i+1}.loc$ from $q_i.loc$. This can be fabricated based on actual conditions.). Algorithm 3 adopts a one-pass approach that can decrease the memory complexity while dealing with quantities of data. Furthermore, to avoid a time span that is too long, Algorithm 3 will check whether the node needs additional time (whether it is reachable with the current situation).

The request time of each point in the trajectory is related to all of its previous points. Figure 3 displays how to change the request time according to Algorithms 2 and 3 when the value of k is set as 5. At the phase of initialization, t_1, t_2, t_3 and t_4 are the request times of fake points while t_0 is the request time of true points. Algorithm 2 forms the trajectory by reorganizing the request time as t_3', t_4', t_0', t_1' and t_2'. If the point of t_3' cannot arrive at that of t_4' within the time interval between t_3' and t_4', then it delays t_4' to t_4''. However, the time interval $t_0' - t_4''$ will be changed due to the delay. It needs to check whether it is accessible from the point of t_4'' to that of t_0'.

Therefore, if the request time of any point ahead of the true point is changed, the request time of the true point may be changed due to Algorithm 3, and it may not be the current time anymore. Even if one point's request time is still the current time, the attacker cannot ensure whether it is a coincidence (e.g., $t_4'' = t_0'$).

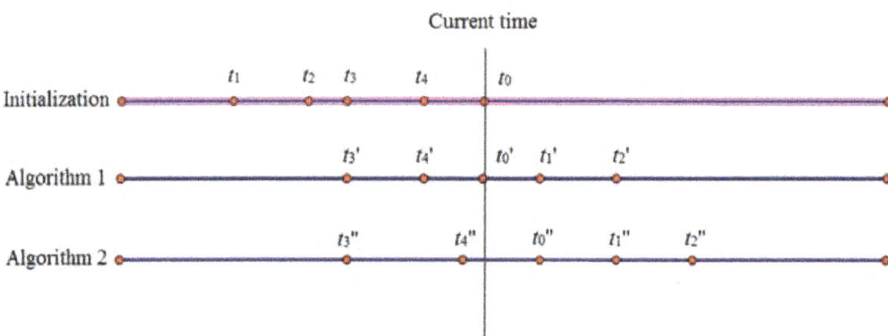

Figure 3. The paradigm of request time change according to Algorithms 2 and 3.

4.4. Inquiry Processing

The anonymity server sends a query request involved in trajectory T, and then the LBS server responds to the request. The LBS server identifies query results by traversing all message nodes in T and then adds the results to a candidate result set W in order. Finally, W is returned to the anonymity server. The final result of this process is $W = \bigcup_{i=0}^{k-1} R(p_i.loc)$.

In the above process, the LBS server traverses all message nodes in T, and W is the union of all message query results after T is received from the anonymity server. When the anonymity server receives W from the LBS server, TTP will find out the answer to the user's original request by selecting that the reply from W whose t before modification matches the user's request time. Finally, the user's real message node will be returned to the user.

5. Privacy Metrics

In this section, it proposes a privacy measure mechanism for evaluating the efficiency of privacy protection in the service system presented. For guaranteeing the availability and effectiveness of the proposed scheme, θ privacy instancing [19] will be adopted.

It uses a series of true user historical location data including position coordinates *loc* and time t, which have been collected by an attacker before the user's current location node is attacked. The attacker can thus probe all message nodes in the credible chain to determine their authenticity according to the latest user location data collected. This is defined as follows:

$$P\{U \mid L_t\} - P\{U\} \leq \theta. \tag{4}$$

In Equation (4), θ denotes the degree of location privacy protection, and, meanwhile, it can be defined as the difference degree of the attack effect between an attacker with background knowledge and someone without background knowledge; $P\{U \mid L_t\}$ denotes the posterior probability that an attacker will infer the user's real location in the current credible chain on the premise that he collects these location data before the moment t; $P\{U\}$ denotes the priori probability that an attacker will infer the user's real location in the current credible chain. Assume that the number of inaccessible nodes excluded by the attacker is α, that the total number of nodes is k, the value of $P\{U \mid L_t\}$ is $1/(k-\alpha)$ and the value of $P\{U\}$ is $1/k$. Consequently, Equation (4) can also be substituted as Equation (5):

$$\frac{1}{k-\alpha} - \frac{1}{k} \leq \theta \tag{5}$$

Assume that the sequence of all location nodes in the credible chain is $p_1, p_2 \ldots p_i \ldots p_k$ and that the latest user location node information collected by the attacker is p_0. The method for calculating the value of θ is described as follows:

Step 1: Judge whether inequality $(p_i.loc - p_0.loc)/v_{0,i} \leq (p_i.t - p_0.t)$ is established. Inequality denotes whether from the location of p_0 the user can arrive at the location of p_i. In this case, proceed to step 2; otherwise, proceed to step 3.

Step 2: Inequality is established, denoting that node p_i may be the user's true node. The value of α is then put into Equation (5) to determine the value of θ.

Step 3: Inequality is not established, denoting that the node p_i is a dummy node. Then, α++ and i++. Proceed to step 1 to continue the calculation.

As is shown in Figure 4, it assumes that k is 4 and that a credible chain $\{p_1, p_2, p_3, p_4\}$ has been constructed by the anonymity server. An attacker collects a series of true user historical location data, and the latest user location data is p_0. First, it determines whether inequality $(p_i.loc - p_0.loc)/v_{0,i} \leq (p_i.t - p_0.t)$ is established according to the above steps in order to calculate the value of θ. As location p_0 cannot arrive at location p_1, p_1 is a dummy one and the value of α is added to 1. It then determines whether location p_0 can arrive at location p_2. As location p_0 cannot arrive at location p_2, the value of α is 2. Subsequently, it determines whether location p_0 can arrive at location p_3. As location p_0 can arrive at location p_3 and location p_3 can arrive at location p_4, p_3 and p_4 may be the true nodes of the user. Finally, it can derive the value of θ as $\frac{1}{4}$.

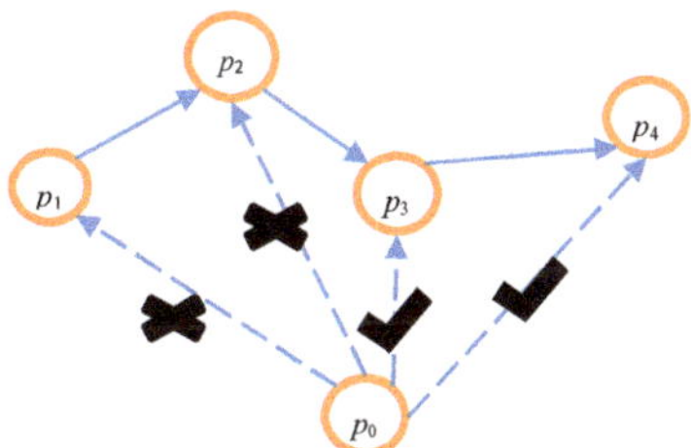

Figure 4. Calculating the value of θ.

6. Experimental Analysis

In this section, the performance of the proposed location privacy-preserving method will be evaluated from three aspects using MATLAB: the degree of anonymity, θ privacy and the quality of service (QoS).

6.1. Degree of Anonymity Analysis

In this paper, the degree of anonymity is determined based on the value of k. It uses a series of data to simulate four weights of influencing factors that can be set as $W_1 = 0.16$, $W_2 = 0.15$, $W_3 = 0.4$, and $W_4 = 0.29$. It samples some location requests from different identities and locations, which are partly shown but not limited in Table 1.

Table 1. The data on the user's environment.

User	Location	Density	Time Interval	Correlation Degree	Associated Number
Student 1	canteen	crowded	morning	high	few
Student 2	hospital	crowded	afternoon	low	numerous
AIDS patient	hospital	crowded	morning	high	numerous
White collar 1	home	sparse	morning	high	numerous
White collar 2	road	sparse	night	irrelevant	numerous
Movie Star	market	extremely crowded	evening	low	numerous
Teacher 1	campus	moderately crowded	afternoon	high	many
Teacher 2	bar	crowded	night	irrelevant	many
Tourist	scenic area	crowded	morning	irrelevant	some

Users have different privacy needs due to their different identities and environments. To obtain a better k value, it analyzes the effects of k_{max} and n on the selection of k using the data shown in Table 1.

Assume that k_{min} is 5 for ease of analysis. According to Figure 5, it can conclude that the value of anonymity degree k should increase as the maximum of the anonymity degree k_{max} increases. This means that the location privacy-preserving method can be used to determine a reasonable k_{max} value to obtain a suitable k according to different anonymity needs while further enhancing the protection of real location data. The value of anonymity degree k should likely decrease or remain unchanged while influencing factor n (the number of levels) grows. When influencing factors cannot be classified specifically, privacy requirements cannot be comprehensively determined or analyzed, and more location messages are needed to complete the anonymity process. It can marginally reduce the number of location messages used in the anonymity process while each influencing factor is accurately classified and measured. It is beneficial to reduce time costs, improve efficiency and protect real location data when forming the credible chain.

Figure 6 displays the relationship between the anonymity degree k, the maximum anonymity degree k_{max} and influencing factors n tri-dimensionally.

As is shown in Figure 6, the value of k will gradually increase as k_{max} gradually increases or as n gradually decreases. This means that more location data must be used in the anonymity process, and, accordingly, communication costs should increase. Thus, the value of k must be set within a reasonable range to limit unnecessary communication overhead; meanwhile, fine-grained classifications of influencing factors and reasonable k_{max} values are needed.

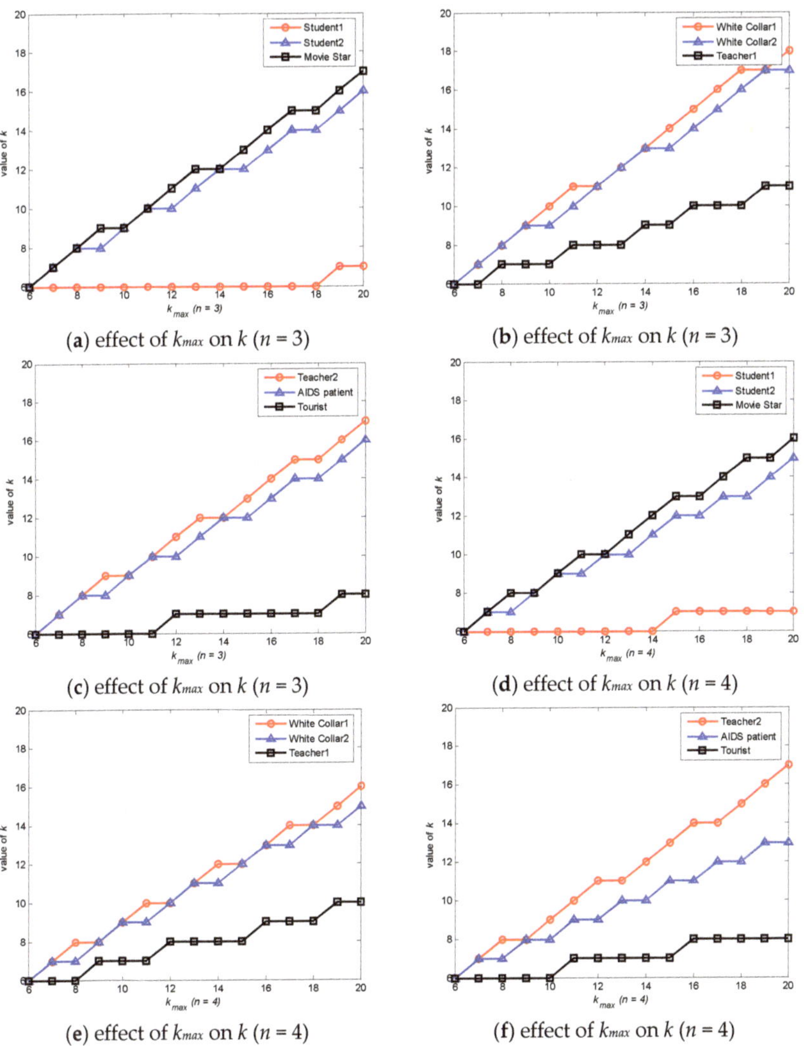

(**a**) effect of k_{max} on k ($n = 3$)

(**b**) effect of k_{max} on k ($n = 3$)

(**c**) effect of k_{max} on k ($n = 3$)

(**d**) effect of k_{max} on k ($n = 4$)

(**e**) effect of k_{max} on k ($n = 4$)

(**f**) effect of k_{max} on k ($n = 4$)

Figure 5. *Cont.*

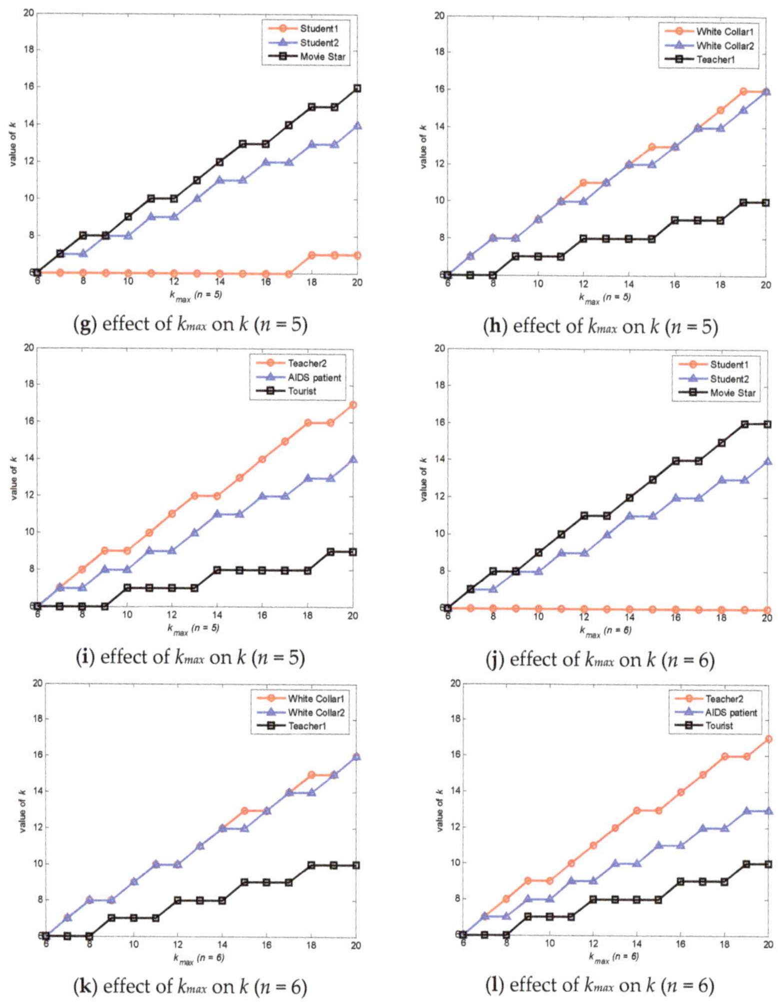

(**g**) effect of k_{max} on k ($n = 5$)

(**h**) effect of k_{max} on k ($n = 5$)

(**i**) effect of k_{max} on k ($n = 5$)

(**j**) effect of k_{max} on k ($n = 6$)

(**k**) effect of k_{max} on k ($n = 6$)

(**l**) effect of k_{max} on k ($n = 6$)

Figure 5. Effect of k_{max} and n on k (**a–l**).

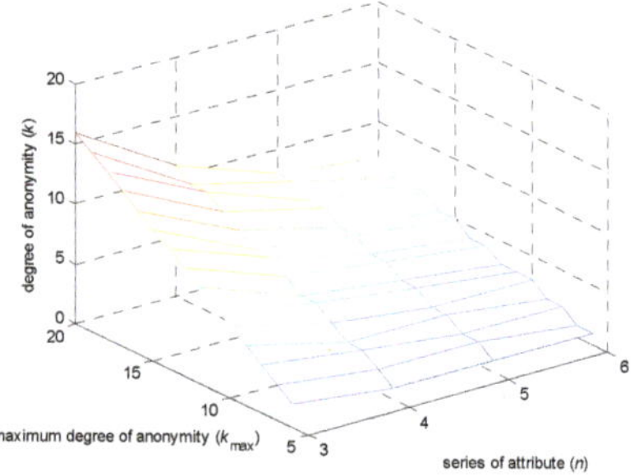

Figure 6. The relationship between k, k_{max} and n.

6.2. θ Privacy Analysis

It uses θ to measure the privacy level. The smaller θ is, the higher the user's location privacy level becomes. In Equation (5), α denotes the number of inaccessible nodes excluded by the attacker according to the user's previous locations. Different values of α in a credible chain denote that the attacker has different background knowledge.

According to Figure 7, as the value of α increases, the value of θ also increases. This shows that when the attacker has more background information, the user's degree of location privacy decreases and protection costs increase, as more fake nodes must be added to the chain. When the value of k increases gradually, the value of θ gradually tends toward 0, which denotes perfect privacy.

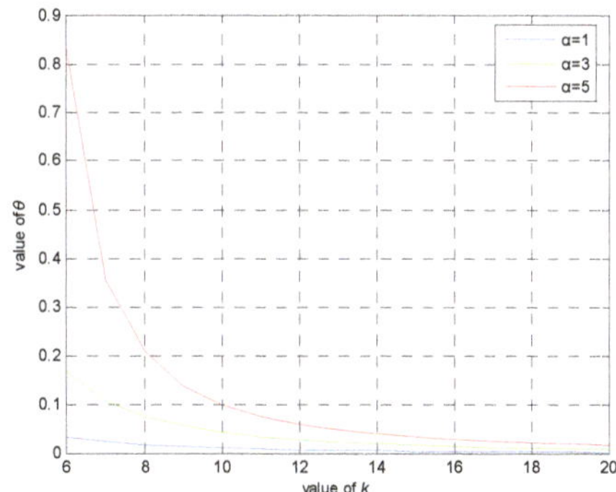

Figure 7. The relationship between k and θ under different background knowledge conditions.

Figure 8 denotes the relationship between k, the number of excluded fake nodes α, and the value of θ privacy.

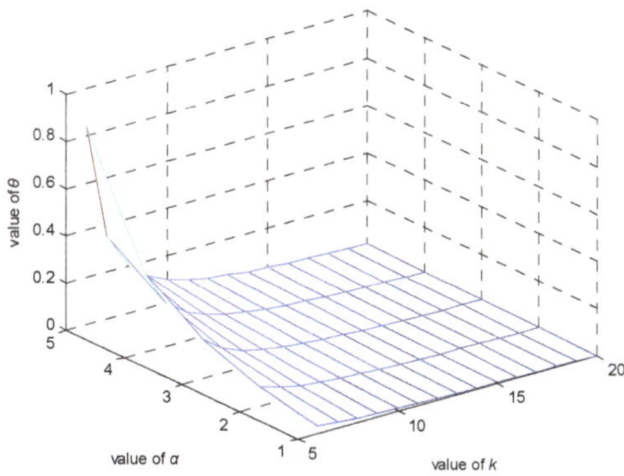

Figure 8. The relationship between k, α, and θ privacy.

According to Figure 8, when the number of excluded fake locations is constant, the larger the value of k is, the smaller the value of θ becomes and the better the degree of privacy protection becomes. When k is constant, the larger the number of excluded fake locations is, the larger the value of θ becomes and the worse the degree of protection becomes.

Assume that the user's current position in the trajectory is x, and the attacker has maximum attack capacities. For example, in Figure 9, A-O-B is a fake trajectory ($k = 8$ in this figure), and P-H-O is the user's true trajectory. Therefore, the attacker holds all of the request messages ahead of O in the true trajectory. It can only ensure that, in A-O-B, for any two adjacent nodes, the previous one to the next one is accessible. Thus, P-O-B is also accessible between any two nodes. It is not certain whether it is accessible from H to any point between A and O (for example, Q). The worst situation is that all points between A and O are excluded. In this case, $\alpha = 4$ ($\alpha = x - 1$).

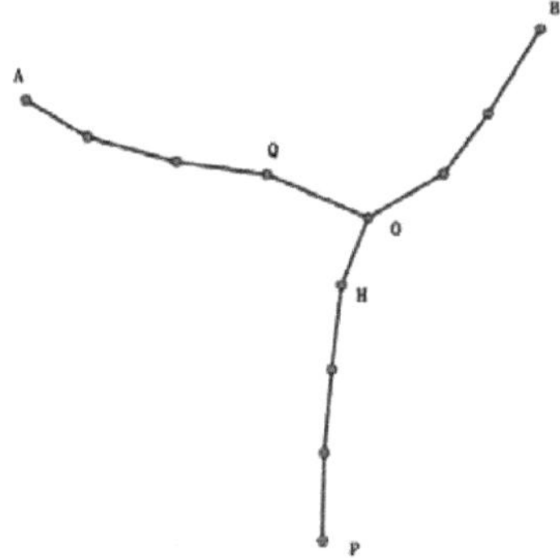

Figure 9. True trajectory P-H-O and fake trajectory A-O-B.

After these analyses, the expected values of α and θ can be figured out, respectively.

Under normal circumstances, the value of α can vary from 0 to $(x - 1)$ with equal probability. Therefore, the following equation regarding α can be derived:

$$E(\alpha) = \sum_{i=0}^{x-1} \frac{i}{x} = \frac{x-1}{2} \tag{6}$$

It can also calculate the expected value of θ by listing all valid combinations of α and x.

The value of x can vary from 1 to k with equal probability. Under this premise, the value of α can vary from 0 to $(x - 1)$ with equal probability. Therefore:

$$\begin{aligned}
E(\theta) &= \frac{1}{k}\sum_{x=1}^{k}\left\{\frac{1}{x}\sum_{\alpha=0}^{x-1}[\theta(\alpha,k)]\right\} \\
&= \frac{1}{k}\sum_{x=1}^{k}\left\{\frac{1}{x}\sum_{\alpha=0}^{x-1}\left[\frac{1}{k-\alpha}-\frac{1}{k}\right]\right\} \\
&= \frac{1}{k}\sum_{x=1}^{k}\left\{\frac{1}{x}\left[H_k - H_{k-x} - \frac{x}{k}\right]\right\} \\
&= \frac{1}{k}\sum_{x=1}^{k}\left\{\frac{H_k-H_{k-x}}{x} - \frac{1}{k}\right\} \\
&= \frac{1}{k}\left\{\sum_{x=1}^{k}\left[\frac{H_k-H_{k-x}}{x}\right] - 1\right\}
\end{aligned} \tag{7}$$

where H_n means the n-th harmonic number ($H_0 := 0$).

With this result, the relationship between θ and k can be depicted in Figure 10, where $k = 1$ means that TTP sends the true point to LBS, while $k = 2$ means that there are only one fake point and one true point in the fake trajectory. These two situations will be excluded in reality. Therefore, it can be concluded that the bigger k is, the smaller $E(\theta)$ and the higher the privacy-preserving level of the user data will be.

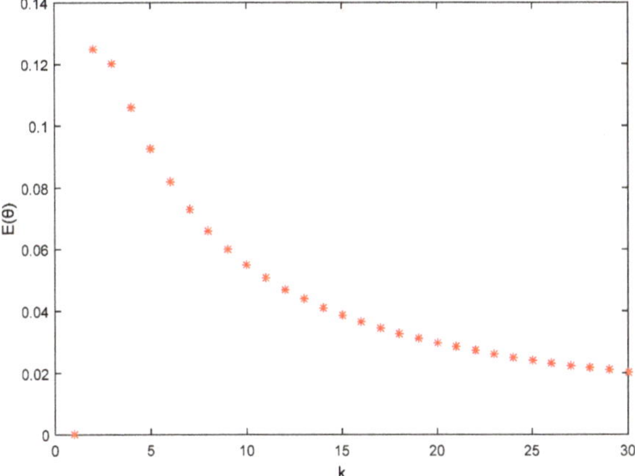

Figure 10. The relationship between k and $E(\theta)$.

6.3. Quality of Service (QoS) Analysis

The following representative anonymous methods are used for comparisons in this section: the quadtree-based [2], Casper [6], service similarity [7] and enhanced-DLS algorithms [8]. Simulation experiments are conducted under the same conditions to compare service accuracy. The service accuracy of an anonymity server can be denoted as $C = \frac{W_{true}}{W}$, denoting the ratio between the valid number of queries and the total value. An increase in C indicates that the accuracy of query results has improved. When $C = 1$, all feedback results are correct. Moreover, 100 random queries are simulated in each algorithm.

According to Figure 11, the proposed method always achieves a service accuracy level of 1 (or 100%), while values achieved by other anonymous methods decline as the value of anonymity degree k increases. This is because the proposed method forms a credible chain based on a user's real location and several fake locations, ensuring that the positional accuracy levels are never reduced. The other methods form ASRs based on k user locations, which decreases positional accuracy levels. From Figure 11, it can be concluded that the proposed method does not suffer from the service accuracy limitations of existing algorithms based on ASRs.

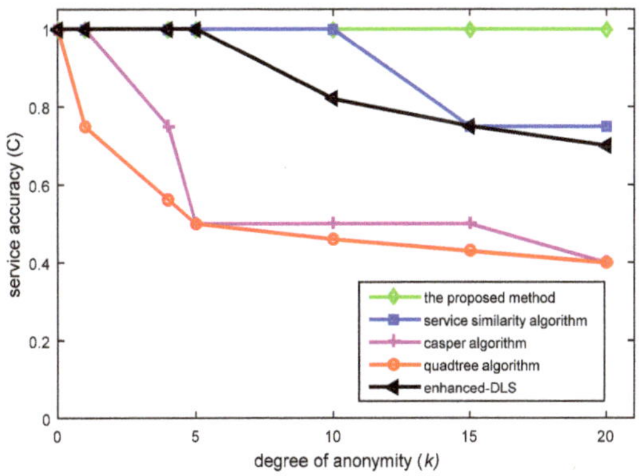

Figure 11. The relationship between k and service accuracy.

7. Conclusions

To address the issue that privacy levels are improved by sacrificing the quality of service in current location privacy-preserving mechanisms, it proposes a location privacy-preserving k-anonymity method based on the credible chain. The method involves utilizing properties of the credible chain and forming a fake trajectory of k location nodes by changing their timing. It also optimizes the value of k and renders it suited to current users' environments and social attributes, thus reducing communication overhead. Furthermore, privacy metrics is suggested by instancing θ privacy. The experimental analysis results show that the proposed method is more effective at addressing contradictions between service accuracy and location privacy. All of the parameters in this paper are extensible and can be changed according to actual requirements. However, the significance of this algorithm is protecting the database and preventing user data from being sold by the LBS providers. If the attacker takes over the LBS and launches a real-time and well-planned attack, the proposed method will degenerate. Future work will involve using big data techniques to analyze and process location data to further improve the effectiveness of location anonymization measures. It is also important to find a suitable method to deal with the real-time and well-planned attack and avoid algorithm degradation. At the same time, the security of third parties and more reasonable k value selection methods also need further investigation.

Acknowledgments: This work was supported by the National Natural Science Foundation of the People's Republic of China (Nos. 61672297, 61572260 and 61373138), the Key Research and Development Program of Jiangsu Province (Social Development Program, No. BE2016185), the Postdoctoral Foundation (Nos. 2015M570468 and 2016T90485), the Sixth Talent Peaks Project of Jiangsu Province (No. DZXX-017), the Jiangsu Natural Science Foundation for Excellent Young Scholars (No. BK20160089), and the Fund of the Jiangsu High Technology Research Key Laboratory for Wireless Sensor Networks (WSNLBZY201516).

Author Contributions: H.W., H.H. and Y.Q. conceived and designed the experiments; H.W. and Y.Q. performed the experiments; Y.W. and M.W. analyzed the data; H.W. and H.H. wrote the paper.

Conflicts of Interest: The authors declare no conflicts of interest.

References

1. Lu, W.; Feng, M.X. Location privacy preservation in big data era: A survey. *J. Softw.* **2014**, *4*, 693–712.
2. Jia, J.; Zhang, F. Non-deterministic k-anonymity algorithm based untrusted third party for location privacy protection in LBS. *Int. J. Secur. Appl.* **2015**, *9*, 387–400. [CrossRef]
3. Sweeney, L. K-anonymity: A model for protecting privacy. *Int. J. Uncertain. Fuzziness Knowl.-Based Syst.* **2002**, *5*, 557–570. [CrossRef]
4. Gruteser, M.; Grunwal, D. Anonymous usage of location-based services through spatial and temporal cloaking. In Proceedings of the 1st International Conference on Mobile Systems, Applications, and Services (MobiSys'03), San Francisco, CA, USA, 5–8 May 2003; pp. 163–168.
5. Gedik, B.; Liu, L. A customizable k-anonymity model for protecting location privacy. In Proceedings of the 25th IEEE International Conference on Distributed Computing Systems (ICDCS'05), Columbus, OH, USA, 6–10 June 2005; pp. 620–629.
6. Bamba, B.; Liu, L.; Pesti, P.; Wang, T. Supporting anonymous location queries in mobile environments with privacy grid. In Proceedings of the 17th International World Wide Web Conference (WWW'08), Beijing, China, 21–25 April 2008; pp. 237–246.
7. Jagwani, P.; Kaushik, S. K-anonymity based on fuzzy spatio-temporal context. In Proceedings of the 2014 IEEE 15th International Conference on Mobile Data Management (MDM'14), Brisbane, Australia, 14–18 July 2014; pp. 15–18.
8. Mokbel, M.F.; Chow, C.Y.; Aref, W.G. Casper: Query processing for location services without compromising privacy. *ACM Trans. Database Syst.* **2016**, *4*, 24–48.
9. Yong, Y.A.; Cheng, L.Y.; Feng, M.J.; Li, X. Location privacy-preserving method of k-anonymous based on service similarity. *J. Commun.* **2014**, *11*, 162–169.

10. Niu, B.; Li, Q.H.; Zhu, X.Y.; Cao, G.H.; Li, H. Achieving k-anonymity in privacy-aware location-based services. In Proceedings of the 33rd IEEE International Conference on Computer Communications (INFOCOM'14), Toronto, ON, Canada, 27 April–2 May 2014; pp. 754–762.

11. Chow, C.Y.; Mokbel, M.F.; Li, X. A peer-to-peer spatial cloaking algorithm for anonymous location-based services. In Proceedings of the 14th Annual ACM International Symposium on Advances in Geographic Information Systems (GIS' 06), Virginia, VA, USA, 10–11 November 2006; pp. 171–178.

12. Chow, C.Y.; Mokbel, M.F.; Li, X. Spatial cloaking for anonymous location-based services in mobile peer-to-peer environments. *Geoinformatica* **2011**, *2*, 351–380. [CrossRef]

13. Zakhary, S.; Radenkovic, M.; Benslimane, A. The quest for location-privacy in opportunistic mobile social networks. In Proceedings of the 2013 IEEE 9th International Conference on Wireless Communications and Mobile Computing (IWCMC'13), Cagliari, Italy, 1–5 July 2013; pp. 667–673.

14. Gheorghita, M.O.; Solanas, A.; Forne, J. Location privacy in chain-based protocols for location-based services. In Proceedings of the 2008 IEEE 3rd International Conference on Digital Telecommunications (ICDT'08), Bucharest, Romania, 29 June–5 July 2008; pp. 64–69.

15. Cao, L.; Sun, Y.; Xu, H. Historical trajectories based location privacy protection query. In Proceedings of the IEEE 11th International Conference on Ubiquitous Intelligence and Computing (ICUIC'14), Ayodya Resort, Bali, Indonesia, 9–12 December 2014; pp. 228–235.

16. Kang, D.; Jung, J.; Mun, J.; Lee, D.; Choi, Y. Efficient and robust user authentication scheme that achieve user anonymity with a Markov chain. *Secur. Commun. Netw.* **2016**, *11*, 1462–1476. [CrossRef]

17. Montazeri, Z.; Houmansadr, A.; Pishro-Nik, H. Achieving Perfect Location Privacy in Markov Models Using Anonymization. 2016. Available online: http://www-unix.ecs.umass.edu/~dgoeckel/zarrin_isita.pdf (accessed on 1 December 2016).

18. Wang, Y.Z.; Xie, L.; Zheng, B.H.; Lee, K.C.K. High utility k-anonymization for social network publishing. *Knowl. Inf. Syst.* **2016**, *3*, 697–725. [CrossRef]

19. Dai, J.Z.; Li, Z.L. A location authentication scheme based on proximity test of location tags. In Proceedings of the 2013 1st International Conference on Information and Network Security (ICINS'13), Beijing, China, 22–24 November 2013; pp. 1–6.

International Journal of
Geo-Information

Article

Proximity-Based Asynchronous Messaging Platform for Location-Based Internet of Things Service

Hyeong gon Jo [1], Tae Yong Son [2], Seol Young Jeong [2] and Soon Ju Kang [1,*

[1]   School of Electronics Engineering, College of IT Engineering, Kyungpook National University,
    80 Daehakro, Bukgu, Daegu 702-701, Korea; tsana@ee.knu.ac.kr
[2]   Center of Self-Organizing Software-Platform, Kyungpook National University, 80 Daehakro,
    Bukgu, Daegu 702-701, Korea; pipikako@gmail.com (T.Y.S.); snowflower@ee.knu.ac.kr (S.Y.J.)
*   Correspondence: sjkang@ee.knu.ac.kr; Tel.: +82-53-950-6604

Academic Editors: Chi-Hua Chen, Kuen-Rong Lo and Wolfgang Kainz
Received: 29 April 2016; Accepted: 8 July 2016; Published: 14 July 2016

Abstract: The Internet of Things (IoT) opens up tremendous opportunities to provide location-based applications. However, despite the services around a user being physically adjacent, common IoT platforms use a centralized structure, like a cloud-computing architecture, which transfers large amounts of data to a central server. This raises problems, such as traffic concentration, long service latency, and high communication cost. In this paper, we propose a physical distance-based asynchronous messaging platform that specializes in processing personalized data and location-based messages. The proposed system disperses traffic using a location-based message-delivery protocol, and has high stability.

Keywords: location-based service; Internet of Things; distributed system architecture

1. Introduction

With the recent development of network and embedded system technologies, interest has grown in the Internet of Things (IoT). For example, a smart home service [1] can check and control appliances in the home by connecting the devices to a network. A smart healthcare service [2] manages the personal medical state by connecting with wearable devices, fitness equipment, and medical equipment. Mobile-asset monitoring [3] provides real-time monitoring and tracking of mobile assets in a factory.

However, despite the devices being physically adjacent, common IoT platforms transfer large amounts of data, generated by multiple devices, to a central server via a global network. This causes problems, such as network congestion due to traffic concentration, service-delay problems caused by multi-hop communication, and a high communication cost because the device is always connected with a centralized server via a global network. In addition, many mobile objects require a complex network protocol, like 6LoWPAN (IPv6 over Low-power Wireless Personal Area Networks) or TCP/IP (Transmission Control Protocol/Internet Protocol), to connect to a network. This is difficult to implement in mobile phones and resource-constrained embedded systems like wearable devices. From the end user's viewpoint, since the personal device information is stored in a central server, privacy violations and security issues [4,5] are possible.

In this paper, to solve the above problems, we propose a physical distance-based asynchronous messaging platform between neighboring nodes. IoT services closely related to people have personalized and localized features, like smart home services and smart healthcare services. Therefore, we define the location-based IoT (LIoT) service, combining IoT services and location information. This specialized platform processes personalized data and location-based messages. The proposed system is composed of (1) a wireless communication proxy that is responsible for direct communication between various mobile nodes; (2) a self-organizing localized IoT messaging hub (SLIM Hub) that makes up the autonomous overlay network; and (3) an ePost-it messaging platform that provides a common format for the services. The proposed system can disperse traffic by delivering messages based on location, increase the system stability via asynchronous message transmission, and solve the privacy issue by using personal or temporary storage rather than unnecessary centralized storage. Especially, our platform can fundamentally prevent personal information leaks by storing private data to personal or temporary storage and providing a service to a user after identifying the user's location.

The remainder of this paper is structured as follows. In Section 2, we introduce related work. Section 3 provides an overview of the LIoT service and the proposed system concept. Section 4 describes the detailed design of the wireless communication proxy responsible for direct communication with a mobile node, and Section 5 presents a detailed design of the ePost-it middleware to provide the LIoT service. Section 6 introduces the implementation of the proposed system and evaluates its performance. Finally, conclusions are drawn in Section 7.

2. Related Work

Real-world location-based applications aim to detect the location of targets in various service domains, such as medical personnel or equipment in a hospital [6,7], smart home management system [8] or stored inventories in a warehouse [9]. There has been research on IoT devices such as Bluetooth low Energy (BLE) sensor module [10] and power consumption issues on IoT devices [11]. EZ [12] is a gateway that supports an efficient asynchronous protocol in IoT. EZ enables the creation of gateways with either C or Java platforms without requiring developers to have any substantial understanding of either relevant protocols or low-level network programming. These research works well-define the characteristics of location-based services (LBS) and IoT. However, they have the aforementioned problems of centralized architecture.

Apple, Inc. has developed the iBeacon [13] service using BLE. A beacon node installed in a fixed location sends a beacon message with its location ID. Then, a mobile node scans this beacon message and determines the location by comparing the signal strength of each beacon. The iBeacon is a localization system without a central server; however, it must go through a service server for real service. NextMe [14] is a phone-based localization system for providing location-based services in IoT. It uses mobile call patterns, which are strongly correlated with co-location patterns. However, it has disadvantages, such as low precision when providing indoor service.

Many researchers have attempted to address the traffic concentration in IoT platforms. Wang and Ranjan discussed the capabilities and limitations of big-data technologies in the fifth installment of "Blue Skies" [15]. The concept of fog computing [16] was introduced to disperse the mass of traffic, based on location. Fog computing can distribute large-scale data and improve the service response time by storing large amounts of data on terminal devices, such as a network router, rather than a central server. However, since fog computing only disperses the data using the network location, regardless of the actual physical location, it is not suitable for location-based services. In addition, privacy issues are still present because it saves all the information in a distributed database on the router.

3. Proposed System Concept

This section provides an overview of LIoT service and the proposed platform concept. First, we introduce the scenario for LIoT under a real environment, and then itemize the characteristics of the proposed LIoT service. The design consideration and the proposed platform concept are followed in the next subsection. Finally, we introduce some protocols suggested in our previous research to realize the proposed platform.

3.1. Overview of the Location-Based Internet of Things Service

Figure 1 is a representation of a location-based IoT service scenario in a hospital environment. A hospital has numerous mobile assets, such as wheelchairs, chemicals, medicine, and hazardous waste materials. The administrator will request typical IoT services, such as (a) monitoring the location and status of equipment; and (b) checking the path of hazardous materials to confirm their safety (shown as green line). In addition, a nurse can request location-based services, such as (c) finding or reserving an available wheelchair near her current position; (d) broadcasting only to the position of the patient who called rather than broadcasting to the entire hospital; or (e) checking the proper medicine by proximity-based direct Device to Device (D2D) communication with a patient's wearable device. Lastly, (f) a user uses the private data such as daily activity or vital signal stored in personal storage.

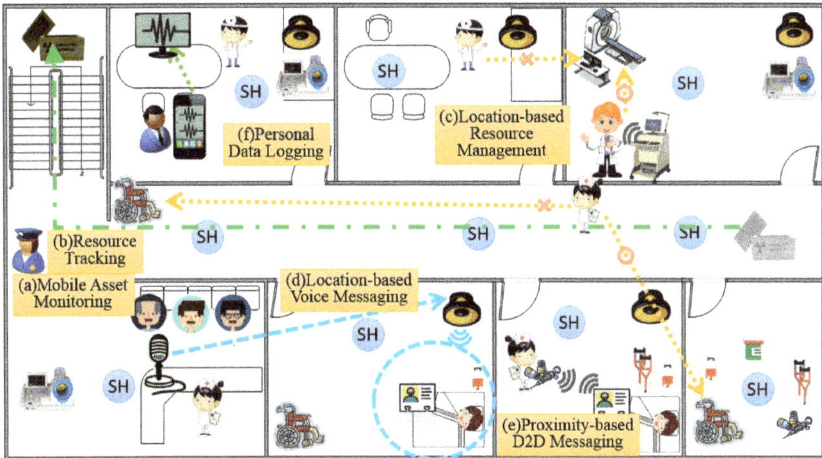

Figure 1. Scenario for Location-based IoT service.

The proposed LIoT service is composed of two types of hardware devices: an end device and a SLIM Hub. The end device is any service device that can identify the user or provide a service. These devices are classified as mobile nodes since they can be moved, even though they may have been installed in a fixed position, e.g., a TV or printer. SLIM Hubs are installed in a unit space such as a room or corridor, representing the installed area and providing a network-access point to the end devices. Thus, an end device can determine its location according to the proximity of a SLIM Hub. In addition, a SLIM Hub is a management device installed at a fixed location to collect assorted information from the end devices in the service area, and provide location-based services based on the proposed messaging middleware.

In this paper, we proposed the LIoT service platform, which is optimized for collecting and delivering localized data by combining the location information and an IoT messaging platform. Devices in our system provide services through proximity-based communication between neighbors, and store personal information to the user's personal device storage. Therefore, the proposed system solves the above-mentioned issues by storing collected personal data to the individual's distributed storage, not the central server, and performing location-based neighbor searching.

3.2. Characteristics of Location-Based IoT Data

LIoT service data has different features than other existing services. Briefly, it has characteristics of both location-based services (LBS) and IoT services. The following are the characteristics of LIoT:

- **Locality**

  Most IoT services have a centralized structure, e.g., cloud-computing architecture. However, LIoT services have location-based characteristics, such as data collection from the required area and service matching based on the current location. Therefore, it is important to combine the location information and service-messaging path.

- **Real-time service**

  In a user-centric service, response time is an important part of the service quality [16]. Particularly, if the service occurs near a user, the response time should be limited to a few seconds to avoid inconvenience for the user. However, in existing centralized structures, the service delay time increases because of the inevitable communication delay when connecting with the server. To solve this problem when using a distributed infrastructure, it is necessary to make the communication distance between the service elements as short as possible.

- **Data sharing and privacy**

  The LIoT service has a variety of service-connection methods between devices: 1:1 connections for text messaging, 1:N for notification messaging, and N:1 for monitoring an area. Hence, the service platform supports a flexible connection method for sharing data. However, the common centralized storage and messaging architecture have a possibility of personal data leak. To prevent such an occurrence, it needs a distributed architecture equipped with a personal storage and a location-based messaging system.

- **Asynchronicity**

  LIoT has steps such as data acquisition, processing, and delivery. In this process, each device connection is asynchronous. Since mobile devices also participate in the service, the LIoT service requires asynchronous transfer capability to improve the communication reliability.

- **Heterogeneity**

  Each service device has many variations, e.g., the type of sensors used for data acquisition, the communication method to be delivered to the service platform, and the format of the data to be distributed. The final stage of the IoT services needs to provide services through generalizing the data obtained from the disparate devices.

- **Small message size, huge data volume [17]**

  The data generated by devices, e.g., sensor values and status, are small. However, the total amount of data becomes huge, since numerous devices participate in the service. To handle these data efficiently, LIoT must have a communication structure suitable for a large number of small messages.

3.3. Design Considerations

To provide a service that matches the characteristics of the LIoT data in Section 3.2, we organized the following requirements and proposed a location-based asynchronous messaging platform.

- **Message-oriented platform**

  To asynchronously transfer a large number of small data messages, message-oriented middleware (MOM) is a suitable messaging platform, since it is designed to rapidly convey large numbers of messages [17]. MOM's publish/subscribe structure has a specialized feature for data sharing between devices. In addition, MOM creates a weak coupling between the mobile nodes and service applications by communicating asynchronously, thereby removing the communication dependency. Thus, it is possible to perform highly reliable communication in an unstable environment, even if traffic congestion occurs.

- **Localization system**

  A localization system is needed to collect data with regional properties and provide location-based services. A localization system recognizes the position of the mobile node and records the location to provide location-based services. In addition, it has a location-based message transmission architecture, rather than a centralized architecture.

- **Protocol gateway**

  The gateway is essential for communicating with mobile nodes that have various protocol types. The protocol gateway provides connection transparency for a mobile node using any protocol by abstracting various communication protocols.

- **Worst-case performance evaluation**

  As described above, an important question in LIoT services is "How long will it take to respond?" The service response time is associated with "How much traffic can a service platform handle simultaneously?" Therefore, in an environment with a large number of messages, the environment should evaluate whether it can respond to the user without the service running out of control.

3.4. Asynchronous Messaging Platform Concept

Figure 2 shows the proposed system's asynchronous-messaging structure. The proposed platform is composed of mobile devices, protocol gateway, and Self-Organized Software platform (SoSp) middleware. Mobile devices are connected to SoSp middleware using various protocols; a mobile device using TCP/IP directly connects to the messaging middleware, and the other protocol devices can communicate to the messaging middleware through the protocol gateway. By just adding a protocol driver into the protocol gateway, the proposed platform can support any communication protocol for mobile devices. A SLIM Hub includes protocol gateway, local storage, and messaging middleware.

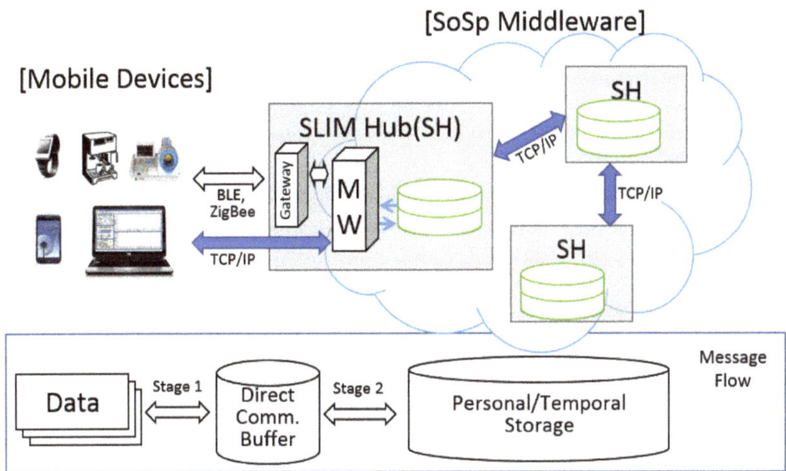

Figure 2. Concept diagram of asynchronous messaging platform.

The messaging flow is composed of the following two communication stages.

- Stage 1: Mobile nodes↔ Protocol gateway

  First, a message generated by a mobile node is transferred to the protocol gateway through direct wireless communication, and stored in the direct communication buffer inside the gateway. In this process, various communication protocols are abstracted by the protocol gateway, so that the message delivery is entrusted to it. For example, for a Bluetooth Low Energy communication, which is usually mounted on a mobile device and has a short maximum transmission unit

(MTU) size of about 20 bytes, we can provide enhanced services without the MTU limitation by abstracting the communication. We named this the 'wireless communication proxy'.

- Stage 2: Message oriented middleware↔ SoSp messaging middleware

After that, the message is transferred to the SLIM Hub (SH), and stored in local storage inside the SH. The stored message will be delivered to personal storage via a preset path, or stored in temporary storage until the SH finds the destination using a location-based neighbor-searching protocol.

3.5. Previous Research for Proposed Platform

Protocols have been suggested in previous research to support the distributed location-based services. By adding the concepts of Section 3.4, based on the localization and discovery protocols, we designed a proximity-based asynchronous-messaging platform.

Location-ID exchange and asynchronous message delivery (LIDx & AMD) [18,19] can provide real-time localization for numerous mobile nodes in a complex and dynamic indoor environment, such as a hospital, warehouse, or museum. Each stationary node installed in a unit space, like a room, periodically sends a beacon message with its location ID. A mobile node can determine its location by selecting the nearest stationary node, which is done by comparing the signal strengths of the beacon messages. This localization protocol uses a simple bidirectional communication between the stationary node and the mobile node, so it guarantees efficient movement of mobile devices.

The location-based service discovery protocol (LSDP) [20] is a resource-discovery protocol. In this protocol, stationary nodes make an overlay network based on the physical-neighbor relationship. A stationary node uses only information about resources within its management range and those of its neighboring stationary nodes. Then, the protocol searches for target resources, using an algorithm similar to graph traversal. This discovery protocol uses only information about the local area and the neighboring stationary nodes, so we can freely look up a resource using distributed-service discovery without a centralized server.

4. Wireless Communication Proxy

A wireless communication proxy is responsible for direct communication with the mobile nodes, using localization and asynchronous message delivery. The proxy is composed of the frontend module (FE) for abstracting the wireless communication, and RFProxy for generalizing the messages from the mobile nodes.

4.1. Structure of Wireless Communication Proxy

Figure 3 shows the elements of the frontend module and the RFProxy agent. The FE supports a variety of wireless-communication protocols using a communication manager. The address book converts an address from a communication address (like a MAC address) to an identification address (like a unique user ID), or reversely. It helps that we can communicate with a mobile device using a unique ID without concerns about the real communication protocol. The network manager and connection manager manage the specific protocol depending on the characteristics of the communication protocol. Particularly, the connection manager manages connect-based protocols, such as Bluetooth, allowing them to communicate with a number of mobile nodes by repeatedly connecting and disconnecting as necessary. The message controller divides a message into a size suitable for the communication protocol and combines the divided message into a single service message once it arrives.

The RFProxy agent abstracts the mobile node information and messages. It communicates with the FE over serial communication, combines the user ID and data from the mobile node, and forwards the message to the appropriate service agent. The message monitor checks the delivery result of the messages transferred asynchronously, and notifies the result to the sender. The message converter abstracts a message by converting a compressed binary to the common format.

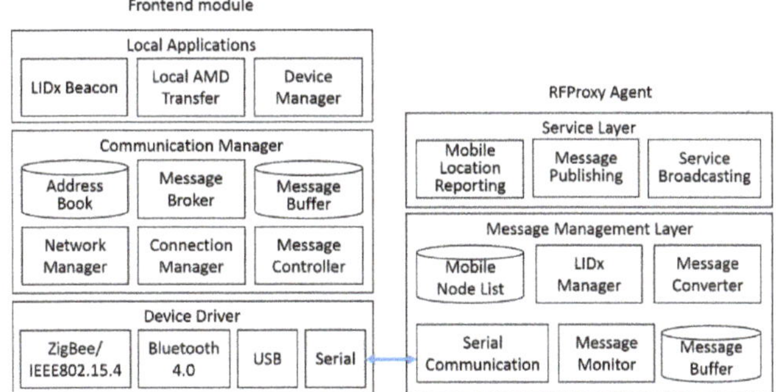

Figure 3. Component diagram of frontend module and RFProxy agent.

4.2. Process of Wireless Communication Proxy

4.2.1. Communication between the Mobile Node and SH

Asynchronous message delivery via direct communication between the mobile node and the frontend module can be classified into two types, depending on the transmission direction, i.e., to or from the mobile node. Moreover, a connection-less protocol, such as IEEE 802.15.4, can easily communicate with a mobile node; however, a connection-based protocol, such as Bluetooth, requires a connection-management mechanism that creates and deletes the communication connection with the mobile node. Following is the communication process for a mobile node and a frontend module using the Bluetooth protocol.

Figure 4 shows how the frontend module sends a message to the mobile node. If the FE receives a message request (1, 2), it switches the Bluetooth mode to a central mode, which can create a connection. When it is connected to the mobile node (3), the FE transmits the message and saves the results (4, 5). Next, the FE forwards the result to RFProxy (7, 8). If the transmission fails because the mobile node has moved, the message is re-transmitted by a location-based protocol, so the message is transferred regardless of the mobile node's movement.

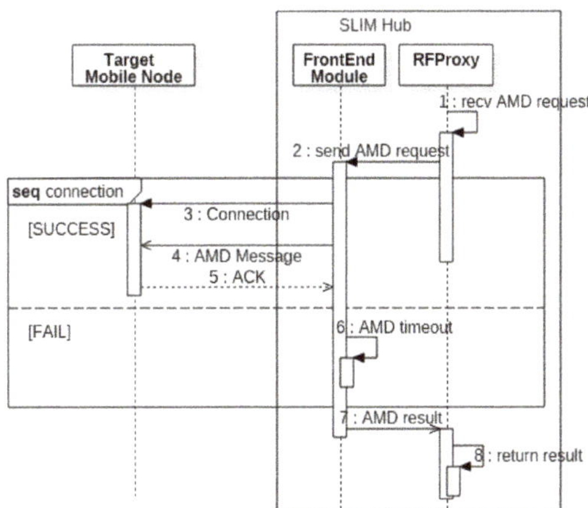

Figure 4. Messaging sequence from SLIM Hub (SH) to target mobile node.

Figure 5 shows the messaging procedure from the mobile node to the SH. When the outgoing message is prepared (1), the transmission starts by advertising that there is a transmission message (3).

If the mobile node knows its location, it advertises with the MAC address of the FE responsible for its position; thus, the SH knows that there is a request from the mobile node (4). Then, the FE initiates a connection (6), communicates with the mobile node (7), and forwards the message to the RFProxy (8). However, the mobile node may not know its current location; therefore, it must request help from the SH to determine which of several FEs it should connect to. When a FE receives an advertising message without a target FE address (9), the RFProxy uses the mobile-node list to determine who is in charge of the mobile node (11). After that, the SH that manages the mobile node communicates with it (12–16).

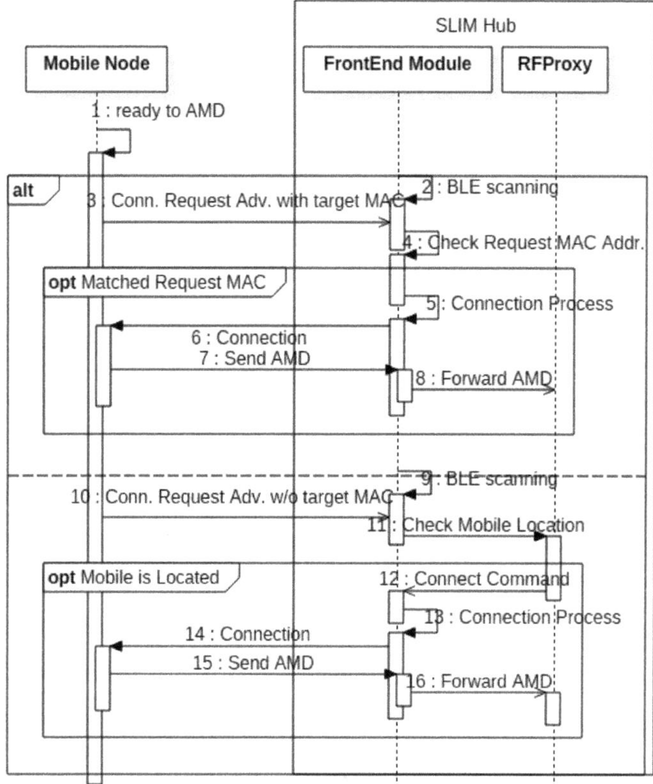

Figure 5. Messaging sequence from source mobile node to SH.

4.2.2. Communication between Mobile Node and Service Agent

Figure 6 is a sequence diagram showing how the service agent and mobile node reconnect when the mobile node changes location. First, the mobile node binds with various service agents in its previous location. Here, "bind" refers to the path for asynchronous message delivery, not the connection binding for synchronous messaging. When the mobile node moves to the new location, the location leader informs the new RFProxy of the movement of the mobile node, with information on the previous RFProxy (2). The new RFProxy receives the service information associated with the mobile node from the previous RFProxy (3, 4), and requests a re-binding to the service agents (6, 8). Now, the service agents and the mobile node can communicate through the RFProxy (10). By running this process as a mobile node changes its location, the proposed system allows the service agents and mobile node to communicate without a central server to relay the connection.

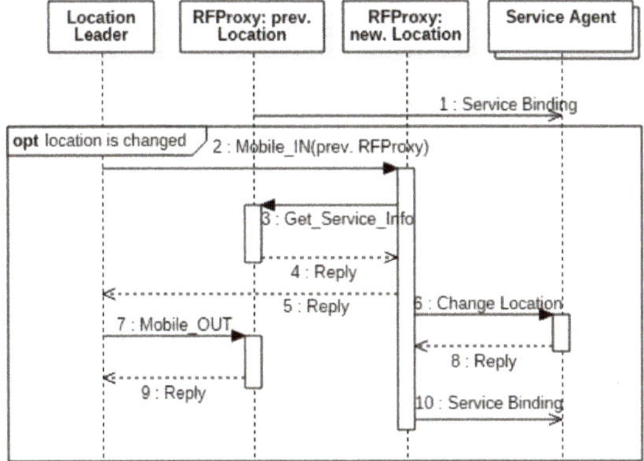

Figure 6. Sequence diagram of service rebinding when a mobile node is moved.

5. ePost-it Middleware

5.1. ePost-it Structure

5.1.1. ePost-it Concept

To implement asynchronous messaging among various devices and services, we propose the ePost-it concept to provide location-based asynchronous messaging. This concept comes from Post-it in the real world, which can be easily separated or attached to the target.

Following are the features of the ePost-it concept, as shown in Figure 7:

- It has a complete message-block type containing a string, encoded binary data, etc.
- It will last as long as the survival time configured in the message.
- It can be sent to all of the users in the region by targeting the area as the destination.
- It can easily be temporarily stored and moved anywhere along the target.

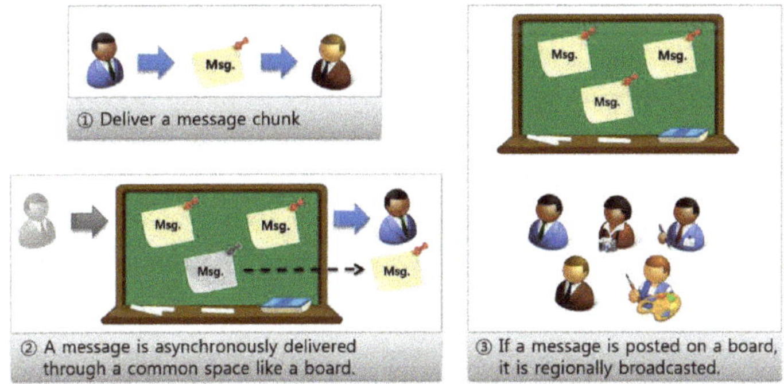

Figure 7. Features of the ePost-it concept.

We defined the common ePost-it data structure in XML format so it can be easily recognized by a variety of devices and services. Figure 8 shows the XML structure of the ePost-it. It has a variety of attributes to process the ePost-it: a callback attribute to obtain the messaging result, and source and destination fields to represent the sender and receiver, respectively. It can select various devices via RFProxy, or select an SH for targeting a region. The contents field can be defined as required for the service. For example, it is possible to include a simple text-based message or an encoded audio file.

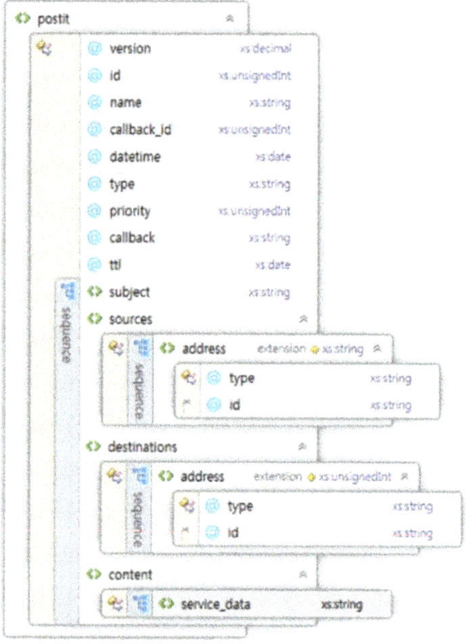

Figure 8. XML schema for ePost-it.

5.1.2. Structure of ePost-it Middleware

In this paper, we propose the ePost-it middleware by adding an ePost-it agent, based on previously developed SoSp middleware [20]. Figure 9 shows the structure of the ePost-it middleware. All SoSp middleware service messages are delivered in ePost-it format through RFProxy or WiFi, and processed by the ePost-it agent. The ePost-it agent analyzes the destination described in the XML document, searches the destination SH to find where the message should be delivered—using the location-based search algorithm [20]—and transfers the message to the messaging agent of the target SH. The messaging agent is responsible for processing and transmitting. First, it stores the message into the local buffer. Then, the message is conveyed directly to the service agents or pushed through a push agent. The push agent transfers the messages to the target according to the device type: it uses RFProxy for small embedded devices and the local push protocol for WiFi devices. If the ePost-it destination is specified as the region, it sends the message to the entire mobile node, to visit the area during the specified time to live (TTL).

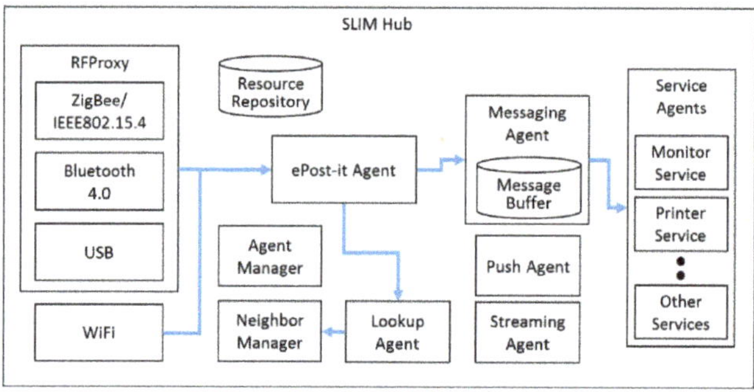

Figure 9. Software structure of ePost-it middleware.

5.2. ePost-it Middleware Process

5.2.1. Message Delivery Process

Figure 10 describes the process of delivering ePost-its. A message generated in the mobile node is transmitted to the ePost-it agent of the source SH (1, 2). Then, the ePost-it agent obtains the target SH information using a lookup engine (3, 4), and transmits the message to the messaging agent in the target-SH managing destination (5). The message stored in the message buffer (6) is transferred through the push agent (7–10). Finally, the ePost-it message is removed, as the message was successfully delivered or its TTL expired.

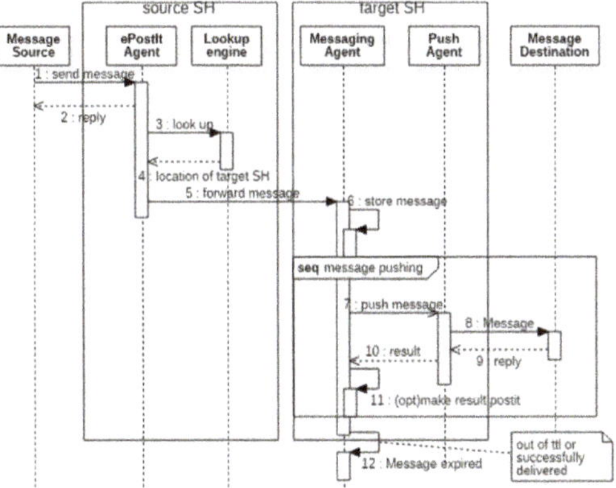

Figure 10. Delivery sequence of an ePost-it message.

5.2.2. Push flow of ePost-it

ePost-it middleware use four types of push methods to deliver a message to a mobile node using various communication protocols. Figure 11 illustrates the process of transmitting an ePost-it to mobile nodes of various kinds. The messaging agent forwards the message to the destinations, depending on the destination type. A message targeted to a service agent is transmitted directly. A message targeted to a mobile node is passed to a push agent, and the push agent transfers the message according to the mobile node's communication type. A global pushing system, such as Google Cloud Messaging (GCM) or Apple Push Notification Service (APNS), transfers the message to a mobile node outside of the SoSp infrastructure. A local pushing protocol transfers it to a mobile node inside the SoSp infrastructure. RFProxy is used for small embedded devices.

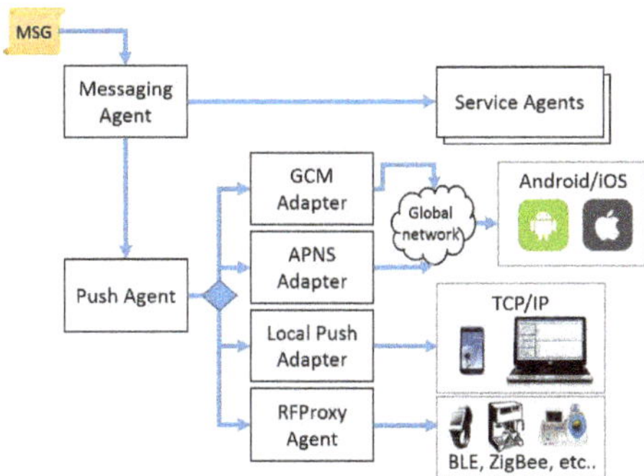

Figure 11. Messaging flow from messaging agent to target agents or devices.

6. Implementation and Evaluation

6.1. Test Environment

Figure 12 shows the hardware used to evaluate the proposed platform. The SLIM Hub acts as a gateway and provides the location-based services (LBS). It includes speakers, LCD, and ethernet access, and is connected to adjacent SHs using TCP/IP. The frontend module is a protocol gateway connected to an SH to abstract the communication with mobile nodes, such as BLE and ZigBee. The mobile tag is a type of mobile node. The tags are used to evaluate the abstraction and management of a connection-based protocol. Finally, to verify the ePost-it middleware performance, a mobile node connected by TCP/IP is simulated on a PC.

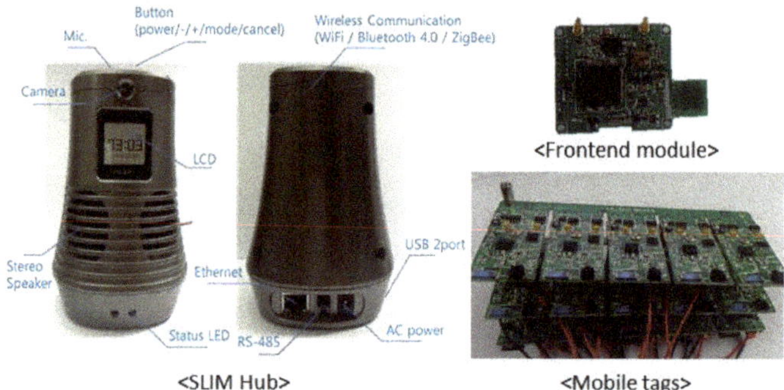

Figure 12. Hardware module used to evaluate the proposed platform.

6.2. Performance of the Frontend Module

To evaluate the FE, we tested the traffic performance. Figure 13 shows the average total transmission time while increasing the number of mobile nodes, each of which sends only one message to the FE. First, a mobile node sends a variable size message to the FE. For a given number of mobile nodes, we measured a total transmission time until all queued requests are processed and calculated an average of 30 runs. As shown in the figure, despite a number of nodes requesting a connection to send a message, the FE can process all the requests sequentially using connection management.

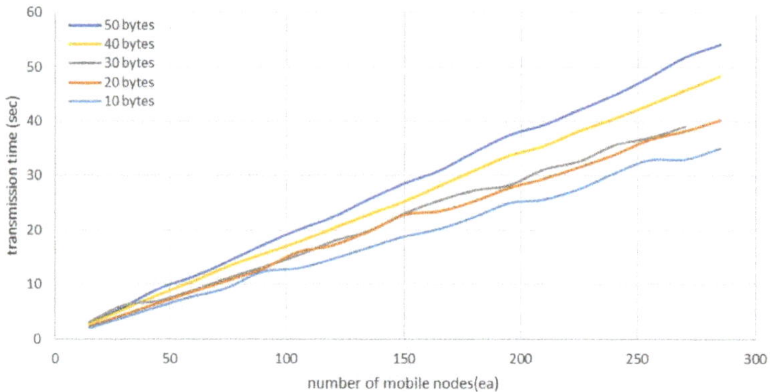

Figure 13. Total transmission time according to the number of mobile nodes.

To estimate the performance with more devices, we measure the time used for messaging during a connection. Figure 14 shows the transmission time for sending a message to the FE according to the message size. The transmission time is composed of the connection time and the data-transmission time, as described in Equation (1). By regression analysis of Figure 14 (95% confidence interval), we determined the constants as $T_{connection} = 110$ ms and $T_{byte} = 1.56$ ms.

$$T_{transmission} = T_{connection} + T_{data\ transmssion} = T_{connection} + T_{byte} \times (number\ of\ bytes) \qquad (1)$$

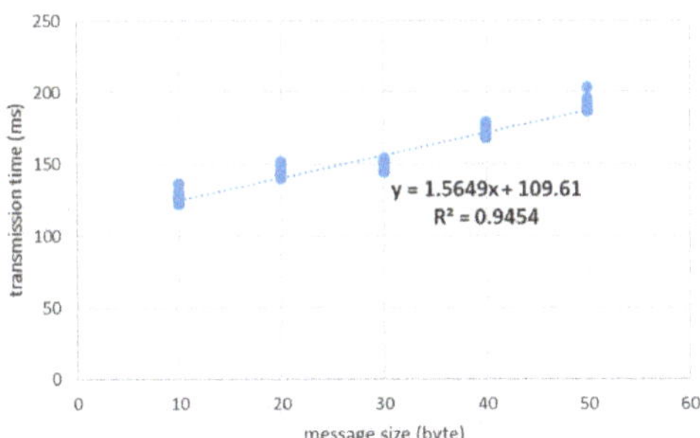

Figure 14. Transmission time per connection according to the message size.

6.3. Performance of the ePost-it Middleware

We created a simple service scenario to verify the performance of the ePost-it. In a conference with a large number of users participating, a presenter wants to distribute the presentation materials to the people in attendance. SHs are installed in each presentation room, and an SH sends the file to the mobile devices by detecting the presence of a user. Figure 15 shows the response time from user's entrance to receiving the message, according to the number of mobile nodes. As shown in the figure, the response time increases slowly as the number of mobile nodes increases. We determined that an SH responds within a few seconds, in spite of numerous mobile nodes.

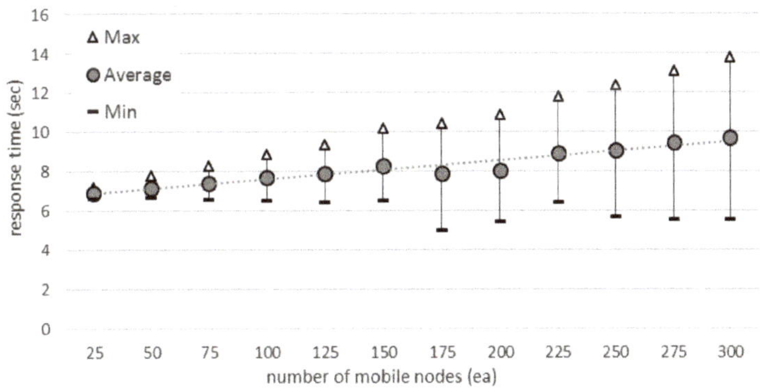

Figure 15. Response time of ePost-it according to the number of mobile nodes.

7. Conclusions

In this paper, we introduced the LIoT service and investigated a proximity-based asynchronous messaging platform specialized in processing personalized data and location-based message. To address traffic congestion and privacy issues, the problems of centralized architectures, the proposed system disperses traffic using a location-based message-delivery protocol, and stores collected personal data to the individual's storage, not the central server. To implement the asynchronous mechanism in IoT, we designed the following components. The direct communication buffer in a FE abstracted the communication protocol and provided a loosely coupled connection between a mobile node and IoT services. The temporary storage held personal data until the data was transferred by the asynchronous-messaging mechanism. The ePost-it middleware supported location-based messaging services. We experimentally evaluated the performance. The results showed that the FE could sequentially process numerous requests, spending a few hundred milliseconds in a connection, and the SH could rapidly provide location-based messaging.

In future work, we will focus on extending the messaging platform to support continuous data streaming, regardless of the device movement; e.g., healthcare applications for a vital-signal streaming system, which requires the transmission and sharing of large waveform data.

Acknowledgments: This work was supported by Institute for Information & communications Technology Promotion (IITP) grant funded by the Korea government (MSIP). (No. 10041145, Self-Organized Software platform (SoSp) for Welfare Devices).

Author Contributions: Hyeong gon Jo designed and performed the experiments, analyzed the data, and wrote the paper; Tae Yong Son designed and performed the experiments; Seol Young Jeong supported the design and analysis; Soon Ju Kang supervised the analysis and edited the manuscripts.

Conflicts of Interest: The authors declare no conflicts of interest.

References

1. Wang, D.; Lo, D.; Bhimani, J.; Sugiura, K. AnyControl—IoT based home appliances monitoring and controlling. In Proceedings of the 2015 IEEE 39th Annual Computer Software and Applications Conference, Taichung, Taiwan, 1–5 July 2015; pp. 487–492.

2. Cecílio, J.; Furtado, P. Middleware solution for healthcare IoT applications. In *Wireless Internet*; Springer International Publishing: Cham, Switzerland, 2015; pp. 53–59.

3. Jeong, S.Y.; Jo, H.G.; Kang, S.J. Fully distributed monitoring architecture supporting multiple trackees and trackers in indoor mobile asset management application. *Sensors* **2014**, *14*, 5702–5724. [CrossRef] [PubMed]

4. Abomhara, M.; Koien, G.M. Security and privacy in the Internet of Things: Current status and open issues. In Proceedings of the 2014 International Conference on Privacy and Security in Mobile Systems (PRISMS), Aalborg, Denmark, 11–14 May 2014; pp. 1–8.

5. Yoon, S.; Park, H.; Yoo, H.S. Security issues on smarthome in IoT environment. In *Computer Science and its Applications*; Springer: Heidelberg, Germany, 2015; pp. 691–696.

6. Kamel Boulos, M.N.; Berry, G. Real-time locating systems (RTLS) in healthcare: A condensed primer. *Int. J. Health Geogr.* **2012**, *11*. [CrossRef] [PubMed]

7. Mun, I.; Kantrowitz, A.; Carmel, P.; Mason, K.; Engels, D. Active RFID system augmented with 2D barcode for asset management in a hospital setting. In Proceedings of the 2007 IEEE International Conference on RFID, Grapevine, TX, USA, 26–28 March 2007; pp. 205–211.

8. Lee, Y.; Hsiao, W.; Huang, C.; Chou, S.T. An integrated cloud-based smart home management system with community hierarchy. *IEEE Trans. Consum. Electron.* **2016**, *62*, 1–9. [CrossRef]

9. Kim, J.Y.; Jeon, B.-W.; Hong, D.G.; Suh, S.-H. A proposition for smart warehouse management system (SWMS) through IoT. *J. Korea Soc. Syst. Eng.* **2015**, *11*, 85–93. [CrossRef]

10. Ryu, D.-H. Development of BLE sensor module based on open source for IoT applications. *J. Korea Inst. Electron. Commun. Sci.* **2015**, *10*, 419–424. [CrossRef]

11. Trappe, W.; Howard, R.; Moore, R.S. Low-Energy security: Limits and opportunities in the internet of things. *IEEE Secur. Priv.* **2015**, *13*, 14–21. [CrossRef]

12. Bromberg, Y.-D.; Morandat, F.; Réveillère, L.; Thomas, G. EZ: Towards efficient asynchronous protocol gateway construction. In *Distributed Applications and Interoperable Systems*; Springer: Heidelberg, Germany, 2013; pp. 169–174.

13. Apple Inc. iBeacon for Developers. Available online: https://developer.apple.com/ibeacon/ (accessed on 15 February 2016).

14. Zhang, D.; Zhao, S.; Yang, L.T.; Chen, M.; Wang, Y.; Liu, H. NextMe: Localization using cellular traces in internet of things. *IEEE Trans. Ind. Inform.* **2015**, *11*, 302–312. [CrossRef]

15. Wang, L.; Ranjan, R. Processing distributed internet of things data in clouds. *IEEE Cloud Comput.* **2015**, *2*, 76–80. [CrossRef]
16. Bonomi, F.; Milito, R.; Natarajan, P.; Zhu, J. Fog computing: A platform for internet of things and analytics. In *Big Data and Internet of Things: A Roadmap for Smart Environments*; Bessis, N., Dobre, C., Eds.; Springer International Publishing: Cham, Switzerland, 2014; pp. 169–186.
17. Dworak, A.; Ehm, F.; Charrue, P.; Sliwinski, W. The new CERN controls middleware. *J. Phys. Conf. Ser.* **2012**, *396*, 012017. [CrossRef]
18. Lee, D.K.; Kim, T.H.; Jeong, S.Y.; Kang, S.J. A three-tier middleware architecture supporting bidirectional location tracking of numerous mobile nodes under legacy WSN environment. *J. Syst. Archit.* **2011**, *57*, 735–748. [CrossRef]
19. Kim, T.H.; Jo, H.G.; Lee, J.S.; Kang, S.J. A Mobile asset tracking system architecture under mobile-stationary co-existing WSNs. *Sensors* **2012**, *12*, 17446–17462. [CrossRef] [PubMed]
20. Jeong, S.Y.; Jo, H.G.; Kang, S.J. Remote service discovery and binding architecture for soft real-time QoS in indoor location-based service. *J. Syst. Archit.* **2014**, *60*, 741–756. [CrossRef]

International Journal of
Geo-Information

Article

Detection of Electronic Anklet Wearers' Groupings throughout Telematics Monitoring

Paulo Lima Machado [1,†], Rafael T. de Sousa Jr. [1,†], Robson de Oliveira Albuquerque [1,†], Luis Javier García Villalba [2,*,†] and Tai-Hoon Kim [3]

[1]   Electrical Engineering Department, Technology College, Universidade de Brasília, Brasilia DF 70910-900, Brazil; paulo.machado@redes.unb.br (P.L.M.); desousa@unb.br (R.T.d.S.J.); robson@redes.unb.br (R.d.O.A.)
[2]   Group of Analysis, Security and Systems (GASS), Department of Software Engineering and Artificial Intelligence (DISIA), Faculty of Computer Science and Engineering, Office 431, Universidad Complutense de Madrid (UCM), Calle Profesor José García Santesmases, 9, Ciudad Universitaria, Madrid 28040, Spain
[3]   Department of Convergence Security, Sungshin Women's University, 249-1 Dongseon-dong 3-ga, Seoul 136-742, Korea; taihoonn@daum.net
*   Correspondence: javiergv@fdi.ucm.es; Tel.: +34-91-394-7638
†   These authors contributed equally to this work.

Academic Editors: Chi-Hua Chen, Kuen-Rong Lo and Wolfgang Kainz
Received: 8 October 2016; Accepted: 5 January 2017; Published: 22 January 2017

Abstract: Ankle bracelets (anklets) imposed by law to track convicted individuals are being used in many countries as an alternative to overloaded prisons. There are many different systems for monitoring individuals wearing such devices, and these electronic anklet monitoring systems commonly detect violations of circulation areas permitted to holders. In spite of being able to monitor individual localization, such systems do not identify grouping activities of the monitored individuals, although this kind of event could represent a real risk of further offenses planned by those individuals. In order to address such a problem and to help monitoring systems to be able to have a proactive approach, this paper proposes sensor data fusion algorithms that are able to identify such groups based on data provided by anklet positioning devices. The results from the proposed algorithms can be applied to support risk assessment in the context of monitoring systems. The processing is performed using geographic points collected by a monitoring center, and as result, it produces a history of groups with their members, timestamps, locations and frequency of meetings. The proposed algorithms are validated in various serial and parallel computing scenarios, and the correspondent results are presented and discussed. The information produced by the proposed algorithms yields to a better characterization of the monitored individuals and can be adapted to support decision-making systems used by authorities that are responsible for planning decisions regarding actions affecting public security.

Keywords: anklet monitoring and tracking; detection algorithms; geoprocessing; Law Enforcement Telecommunications Systems (LETS); sensor data fusion

---

1. Introduction

The use of electronic anklets by investigated and convicted persons has been applied by the Brazilian enforcement authorities to try to reduce mass incarceration in the country. According to December 2013 data from the National Penitentiary Department (DEPEN - Departamento Penitenciário Nacional) of the Ministry of Justice, Brazil has one of the largest prison populations in the world, with 581,507 inmates [1]. The number of people incarcerated increased 52% in the 2005–2013 period. At the same time, there is a growing deficit of prison capacity (Figure 1), resulting in overloaded prisons. In Figure 1, the presented information was extracted from the Penitentiary Information

Integrated System (INFOPEN - Sistema Integrado de Informação Penitenciária) from DEPEN, last updated in July 2014.

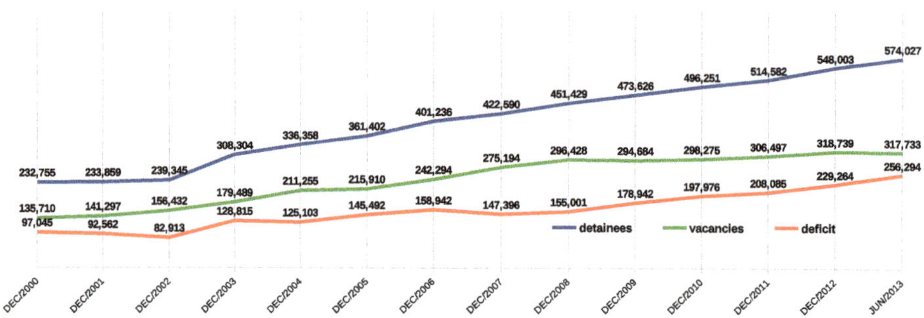

Figure 1. Evolution of the Brazilian incarcerated population; source: authors adapted from [1].

Sentence serving effectiveness inside current overloaded prisons has been stressed by indoctrinators, who claim that such prisons, together with their management methods, are a failure as a means of rehabilitating offenders. Moreover, these indoctrinators argue that this institution has not been proven for behavior rehabilitation, but serves on the contrary as a "true school for criminals".

In principle, prison sentence serving in Brazil must follow well-defined parameters regarding the respect to and the dignity of the prisoner, a requirement of the entire prison community, thus respecting the limitations arising from the sentence, as well as social, economic and cultural rights. Therefore, the possibility that the convict regains her or his dignity through social interaction is one of the goals that guides all sentence serving schemes and, consequently, law enforcement. It seems that a remedy contributing to rehabilitation is to re-socialize a convict, asking the State to rigorously perform the monitoring of this process.

In this sense, the use of convict surveillance by telematics means has proven to be a viable alternative for monitoring sentence servings, thus leading to the development of innovations in the control of individuals who violate criminal laws. One such alternative is the monitoring of convicts who are required to wear electronic anklets that integrate GPS sensors, a form of surveillance that proved effective both in the United States and Europe [2]. Thus, criminal justice systems use GPS devices to monitor offenders, individuals that are out of prison, but forced by law to wear anklets that report their locations to monitoring agencies [3]. This electronic monitoring of convicts may be considered an effective means of social reintegration, which can be gradual since the individual monitoring can be adapted to different sentence serving regimes, as for instance the closed, semi-open and open regimes in the Brazilian legal system.

The use of electronic monitoring brings a breakthrough to the criminal enforcement system. First, from the social point of view, it provides better reintegration for the rehabilitating convict into a society that otherwise is not prepared for dialog and much less prone to assist convicted individuals in the social reintegration process. Furthermore, electronic monitoring answers the issue of the intimate life privacy of the monitored persons and their families. Taking into account that currently "35% of the prison population in Brazil is made up of pre-trial detainees and 30% of inmates were sentenced for committing crimes without violence or serious threat" [4], electronic monitoring is presented as a key to evolving the penitentiary system.

Another advantage is that electronic monitoring saves public money expenses regarding the prison system, especially considering the issue of the operational effort and cost of surveillance. A succeeding convict monitoring system can reduce, to a great extent, the number of gradual steps to freedom that are currently employed in prison regime progression, including temporary leaves. Similar savings can be obtained from other applications for convict monitoring without direct supervision.

The architectures used in monitoring systems with electronic anklets follow, in general, the following workflow: geodesic coordinates collected by the anklet devices are sent through a GSM/General Packet Radio Service (GPRS) data transmission network via mobile network operators, thus pushing the data to the monitoring center. This latter processes geographic data from the anklets and issues reports and alerts for the relevant authorities to take action in accordance with the policies of each state. The most common monitoring information output is the indication of a tracked offender entering forbidden geographical areas.

Following, we briefly describe the main elements that are part of the structure used for an anklet monitoring system, since these elements are necessary for understanding the system operation and its respective data stream:

- GPS:

  Global Navigation Satellite Systems (GNSS) comprise constellations of satellites that by transmitting signals to a receiver, make it possible to determine its coordinates. These signals, which are transmitted at specific frequencies, possess peculiar characteristics that allow their identification by receivers, characterizing what is called GNSS observables [5]. The GNSS adopted for this paper presentation is the North American Global Positioning System (GPS). Through this system, electronic devices constantly receive signals from satellites and determine their distance from these satellites by calculating the time elapsed for receiving the signal and their speed. Given the distances between the devices and the satellites, the device calculates its relative position on the globe and generates a pair of numbers that identifies its location (coordinates), observing a specific georeferencing system.

- GSM/GPRS:

  The Global System for Mobile (GSM) standard is a digital communications system that allows data to be moved both synchronously and asynchronously and also preserves the GSM Short Message Service (SMS) existing in previous systems [6]. General Packet Radio Service (GPRS) is considered an intermediary between GSM and 3G cellular networks, offering data transmission via a GSM network in the range of 9.6 Kbits–115 Kbits. Furthermore, GPRS technology supports telephone calls and data transmission at the same time, thus allowing for example a GPRS mobile phone user to make calls and receive e-mail messages, simultaneously. GPRS reserves radio resources only when there are data to send and reduces reliance on traditional circuit-switched network elements, then enabling IP protocol data transmission over GSM [6].

- The monitored device:

  The device comprises a box with an electronic circuit equipped with a GPS module for geolocation. It also has slots for one or more mobile network connection SIM cards for pushing data via GSM/GPRS. In order to maintain a continuous operation, the device has a built-in battery that must be recharged periodically by the user via a charger supplied with the equipment. A survey carried out among electronic anklet suppliers shows that most devices have the following characteristics: quad-band GSM/GPRS 850/900/1800/1900 MHz, GPS signal reception from at least 20 satellites, the ability to operate with one or more mobile operators (multiple SIM card slots), sufficient memory to accumulate at least the last 24 h of trajectory in case of off-line communications, at least 24 h of battery life, a sensor and warning indicator for low battery events, a sensor and warning about the physical violation of the device, a radio jamming detection sensor and data communication encryption.

- The monitoring central system:

  The most common information output from an anklet monitoring system is an indication of a tracked offender entering restricted areas. Thus, its central system must deploy storage and computing resources able to capture data from the monitoring network, organize this data, perform calculations in maps, register forbidden areas for individuals and support the functions of authentication, authorization and auditing.

Moreover, this paper considers the possibility that the monitoring center can also provide additional services related to data on the formation of groups of monitored individuals, based on proximity detection regarding the coordinates provided by the anklets. In addition to showing the groups and their location, it is also possible to consolidate information on the time elapsed during which each group remains together, the number of individuals, as well as the frequency and time of the meetings. Such information has the potential of contributing to risk analysis that includes preventive actions by law enforcement agencies.

Law Enforcement Telecommunications Systems (LETS) should take into account the actual risk posed by specific groups, taking into consideration factors, such as the danger level posed by their elements and the types of offenses committed by each of them, among others factors [7]. Therefore, it is important to design algorithms that provide data in order to corroborate risk assessments and decision-making in this context. The objective is to thereby issue alerts informing of probable riot formation, preparation for criminal activities, among other suspicious activities.

The core contribution of this paper is to design a set of articulated algorithms, providing a systemic model able to process data from the monitoring network in order to: (1) verify proximities (detection of pairs); (2) group devices that are in proximity with each other into clusters (detection of groups); and (3) record groupings' duration and the average number of grouped elements (detection of risks). Additional contributions are described regarding the implementation and performance aspects of these algorithms. It is interesting to point that such algorithms are applicable in other situations, e.g., monitoring animal groupings in forests.

The remainder of this article is organized as follows: Section 2 discusses related papers. Section 3 describes the problem of grouping detection and introduces the systemic model of the proposed solution. Section 4 provides an analysis of the systemic model and the results from the algorithms in a simulated environment. Section 5 concludes this paper and presents possible further research.

2. Related Works

This paper subject pertains to the general domain of multi-sensor data fusion ([8,9]), but is more specifically related to the works presented below.

Papers addressing geographic point processing and cluster identification are generally based on the search for the concentration of points by analyzing their distribution. However, they do not take into account the specific need of identifying individuals gathered at points that are within a minimum distance, which characterizes a meeting. Without such consideration, a possibly detected concentration of points can refer to points separated by distances to the order of kilometers and not just a few meters, which is inconsistent with the concept of a meeting or gathering of monitored people.

Liu et al. [10] addresses algorithms that identify clusters of objects classified into categories, considering purely geographical aspects or other associated attributes. It mainly discusses the "Density-Based Spatial Clustering" (DBSC) algorithm that identifies clusters by using both spatial proximity and attribute similarity. DBSC involves building proximity relationships between points obtained through Delaunay triangulation [11]. In order to obtain the triangles formed by the points in the proximity required by the algorithm, the distance between the points must be previously calculated, but without considering a time frame restriction that we consider in the present paper. The cited paper assumes that the geographic points are static and do not consider any displacement and transformation of clusters over time, changing characteristics that are also considered in our view of groups.

Carlino [12] argues about the influence of the physical proximity of research and development (R&D) laboratories on the impact of knowledge in their area of concentration. For that effect, it compares the location of laboratories in the U.S. territory with patent registrations in the same area, showing their connection. Then, it approaches a way to measure the extent of the spatial concentration of activities of laboratories and defines the cluster formed by neighboring laboratories considering a circle around each location point with an initial radius of a quarter mile. It then lists the number of points within the circle. As a result, many circles overlap, thereby forming the cluster to be analyzed and compared with the registration of patents. It also considers static points in relation to the addresses of laboratories. In the problem presented, there is no need to analyze a change in the cluster over time. Additionally, the cluster area is obtained by delimiting circles in the geographic space applied to all points, which in the article is fixed approximately at 1000 without a perspective of growth.

The DBSCAN algorithm proposed by Louhichi et al. [13] seeks to identify clusters with different kinds of geographical objects (points, polygons, lines, etc.). Each adjacent group in a given radius must contain at least a minimum number of points, i.e., its density surpasses a given threshold, which makes clear that point to point processing is performed by using the relationship of distance between points similarly to the present paper. The cited paper proposes estimating the distance value in order to distinguish the idea of the concentration of points from the idea of scattered points outside the concentration (noise). However, in our present paper, this value is not necessary since we use the GPS precision (accuracy).

The above papers are not in the field of LETS and do not meet the requirements of the problem addressed in the present article, namely: (1) they do not consider the evolution of the group over time by identifying the duration of the concentration of points and the size (number of points) of the group; (2) they do not have a time frame processing threshold and cluster identification; and (3) some algorithms do not impose a minimum distance limit between points in the clusters.

Morreale [14] proposes a design for Wireless Network Information and Identification System sensor (WINS Id) where a large volume of geographically distributed sensor temporal data is collected, stored and presented in real time. This article does not compare the results of real-time processing with previous results showing some evolution for analysis. A basic difference between the monitoring architecture for electronic anklets and the sensor network architecture is the fact that in the first case, there is no daisy-chaining or concentration of data traffic nodes within the network, since in anklet monitoring, the data are sent directly to the monitoring center responsible for processing the data as a whole. This design meets the simplicity of anklet devices designed to connect via GSM/GPRS networks.

Another related field for this paper is the study of data mining techniques on the collected and stored data to knowledge discovery, such as Zhu [15]. In this case, variations on the number of identified groups, number of group elements, frequency, etc., can be processed by the DTW technique for raising monitored abnormal behaving individuals as a whole. It proposes a single system to record offender events with a focus on mobile devices where the current location of the device is used to identify the geographic area where the event occurred. The geographic coordinates are gathered from devices, such as smart phones or tablets, while in the present article, we refer to electronic anklets with less processing power. The cited paper proposes as future work applying data mining techniques on the records in order to establish preventive measures against crimes. In this sense, we consider that integrating a system as proposed by Jakkhupan and Klaypaksee [16] with monitoring by anklets could in certain circumstances accelerate misbehavior detection by identifying suspects present in the crime area at the time that a crime occurred.

Using data mining techniques, Sathyadevan [17] proposes an approach to predict crimes by geographical areas. The processing flow comprises data collection, classification, pattern identification, prediction and visualization. Among the sources of the data, the paper cites "web sites, news sites, blogs, social media, RSS feeds etc.", and the unstructured data are stored in MongoDB. The structured records and groups identified by our technique discussed in the present article could be added to enrich the predictive analysis of the occurrence of crimes. The cited paper demonstrates the development of a mobile application for criminal case records, removing the need for citizens to go to a police agency to fill out bureaucratic forms. Thus, in addition to increasing the number of recorded incidents (many are not registered because of the bureaucracy), it also reduces possible errors in filling, providing for instance the correct indication of the place of occurrence.

The intersection of data from electronic anklets, as given by the proposal in the present paper, with records of occurrences suggested by Oduor [18] could provide better support for the investigations of those cases. It proposes a monitoring architecture for electronic anklets with a topology that considers interim autonomous agents between devices and a center. Agents are dynamic software components that provide collaborative operation services. Using these agents, the system can make decentralized decisions, streamlining the alerting process. Park [19] cites as an example the various levels of warnings about the proximity of a sex offender and monitored children. However, the work provides no details about the infrastructure and the location of these agents and how to connect to the devices and the control panel.

On the other hand, Urbano and Dettki [20] address the issues of creating and maintaining a database in PostgreSQL with the PostGIS extension, which stores geographic data transmitted by sensors located in Italy. It describes the steps for creating the database and the necessary tables for geographic data demonstration and storage. The present paper complements such analysis with more details on the database and implementation requirements in order to validate the algorithms presented hereafter.

Given the need to process the geographic points within a specific time window, even with a large amount of geographic coordinates in the collected sample, it becomes relevant to adopt algorithms that can be parallelized, especially as regards the identification of pairs of points in proximity. Therefore, it is interesting to cite Ding and Densham [21], who present some options addressing the possible division of a geographical space for processing parallelization.

3. Description of the Problem and Systemic Model

Satellite-based device tracking systems consist of several integrated technologies to track rehabilitating convicts in open and semi-open serving regimes and under house arrest. Associated with the joint actions by the civil and military police, these systems allow efficient law enforcement through a monitoring center, which transmits the alarms to the police stations nearest to the locality where an irregular event is detected by monitoring devices.

Several companies offer electronic monitoring solutions through anklets in Brazil and the world. As a basic functionality; they use GPS geolocation equipment and send location data through mobile phone networks, identifying zone violations in the form of inclusion (areas the monitored convict cannot leave) or exclusion (areas the monitored convict cannot enter). The monitoring center is responsible for processing the location points and sending alerts to the appropriate authorities in the case of such violations.

3.1. The Problem

From the point of view of law enforcement monitoring, the concentration of monitored devices in a geographical area does not necessarily indicate a grouping of individuals in a meeting. For example, considering a concentration of monitored points in a geographical space area where the shortest distance found between observed points is 1 km, one cannot immediately deduce that the monitored subjects are actually in a meeting, although the observed concentration is even visually observed in a map. In fact, for two or more monitored individuals to be considered together in a group, the distance between the points representing these individuals must be less than a certain proximity threshold. Precision on the concept of proximity is given hereafter in Section 3.1.1.

The algorithms proposed in this paper are required to perform the processing of geographic points to identify groupings of monitored subjects considering such a threshold distance, in addition to updating a database with additional data, such as the duration of group formation and its number of elements.

Furthermore, the algorithms' steps should be performed in a period of time that does not exceed an established processing window due to law enforcement requirements regarding the freshness of monitored information. This window is parameterized and arbitrarily set at one minute without any prejudice to the obtained results. Moreover, it is interesting to comment that this value is also a performance threshold for our algorithms, because if this time window is exceeded, there is a risk of accumulating the processing of successive actualization windows, possibly overloading the processing and storage sub-system or leading to information loss.

Another important issue is that, for a system to monitor rehabilitating criminals, which implies public security concerns, the calculation of the real risk posed by a group involves much more factors, including the level of danger of grouped individuals and the types of offenses previously committed by each of them, among other factors. Therefore, our processing algorithms shall provide data to support risk analysis, not being ultimately responsible for the analysis itself. This is an important consideration before addressing the concept of proximity adopted throughout the remainder of this paper.

3.1.1. Definition of Proximity

For two or more monitored individuals to be considered together in a group, there must be a minimum distance established between the points representing them. The algorithms proposed in this paper are functionally specified to consider this distance for processing geographical points in order to identify groupings of monitored convicts.

The minimum distance that characterizes a meeting, which is used as a threshold in the processing, must take into account the margin of error (ε) inherent to GPS equipment (Figure 2).

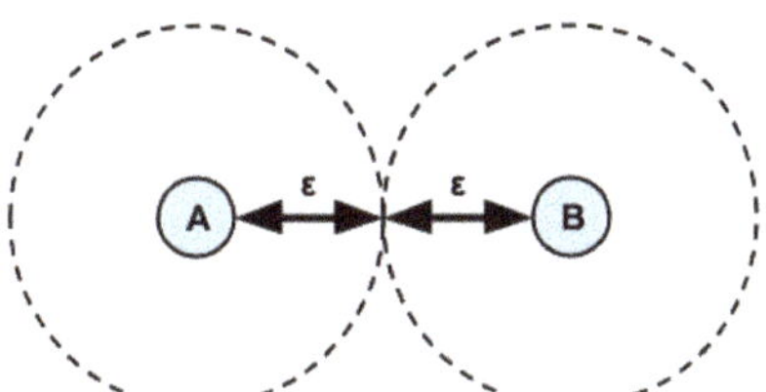

Figure 2. Minimum distance between two points to characterize proximity. Source: the authors.

According to the National Satellite Test Bed/Wide Area Augmentation System (NSTB/WAAS) T&E Team [22], the accuracy of GPS devices is slightly smaller than 10 m. Thus, considering this margin of error, Equation (1) is applied to set the meeting distance threshold (lr). In other words, in practice, for calculation purposes, any two points separated by a distance under 20 m will be considered monitored subjects in proximity.

$$lr = 2 \cdot |\varepsilon_{min}| \tag{1}$$

3.1.2. Duration of a Possibly Detected Group

Detection of groups is performed considering not only the grouping of points in space at a given moment, but also the evolution of this group over time. Thus, indicators, such as group duration and average number of elements, are pieces of information that can be generated by comparing and identified groups in each sample points sent by the devices. Our proposed algorithms shall then provide for the processing of this information.

By maintaining a base of active and inactive groups updated at every sample processing, other information can be easily extracted such as the frequency and time that each group meets. This information supplements the analysis showing any real risk of imminent criminal action or a continued criminal relationship.

Figure 3a,b illustrates the measurement of group duration, respectively in situations where people are standing or moving. During the interval for computing points proximity, *t0* is a specific time when there is not enough proximity between points to consider them as being grouped. At time *t1*, with the points coming close to each other, they are considered to be part of a group that at minimum has two member points. During the following processing times (*t2* and *t3*), the same points still remain within the proximity range. At time *t4*, the two points separate from each other. The system will compute *t1* as the start date and time of the group meeting and *t3* the end of this meeting. Such data comprise the duration of the group existence. This same reasoning shall be applied both to stationary (Figure 3a) and mobile (Figure 3b) points in proximity.

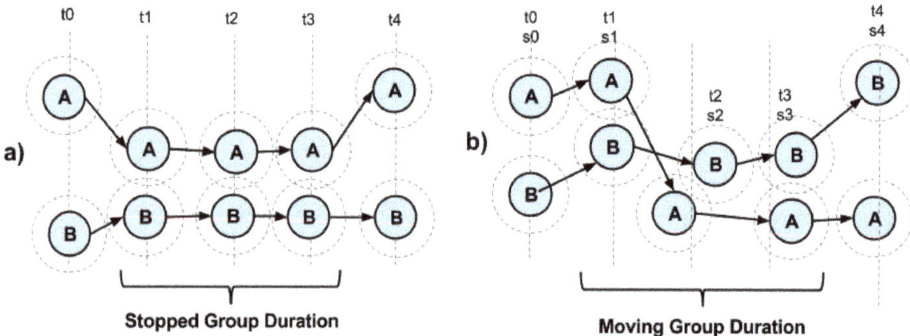

Figure 3. (**a**) Anklet monitoring system with stopped group detection; (**b**) Anklet monitoring system with moving group detection. Source: the authors.

When two monitored individuals intersect in some location, for instance an avenue, their physical proximity may be detected by these calculations, but it does not necessarily mean a grouping of monitored individuals. In order to avoid such situations defined as "false positive", groups whose duration is less than a predetermined value should be discarded. Initially, this variable is set to a minimum of 5 min. In other words, considering minute to minute samples, when the same group is identified in the processing of five consecutive samples, those points in proximity will be considered as an effectively detected group.

3.1.3. Number of Elements in a Detected Group

The number of elements that are part of a grouping influences the evaluation of risk. For example, groups of five elements can pose a greater risk than groups of two elements, as this situation may represent a more severe and organized offense through the division of activities between group members. Hence, providing the number of individuals in a group at the end of the processing is important to support decision-making.

We should also consider that during the existence of a group, its number of individuals may increase or reduce, variations that can be detected by computing their proximity. These variations do not disqualify the group. Thus, we consider the average value of the number of individuals in the group during its existence, an indicator that allows us to consider proportionality in possible comparisons among groups.

However, a variation in the number members of a group can impact the comparison of this group with previously detected groups. This brings the question of how to accurately establish that a previously detected group that had, say, 10 individuals is for the most part the same as one that now has 12 elements: how many members the two groups have in common that yield the conclusion that one group is indeed a reduced or an expanded incarnation of the other. In order to consider this sort of recurrence of a group, we include in the routine that performs comparisons of groups a variable called "commonality", which corresponds to the number of individuals common to both groups (current and former) divided by the total amount of former group elements, expressed as a percentage. If the commonality among two groups is equal to or greater than a commonality threshold, which is initially set to 50%, we consider that they are the same group, and in this case, an attribute containing the average amount of these group members is properly updated. Otherwise, the group under analysis is considered a new group to be remembered.

3.1.4. Time Limits for Running the Algorithms

Anklet devices are configured to periodically send their geographical coordinates or points, typically every 1 min approximately, although this time is usually configurable. Thus, the algorithms to identify groups and gather associated data must run in less time than this whole period boundary, i.e., before the next set of coordinates arrives for new calculation. Moreover, this limit is a performance threshold, because, if this limit is exceeded, there is a risk of accumulating tasks, or computing threads with the processing of the previous set, or overloading the equipment responsible for processing, or loosing information. Therefore, the whole algorithm must run in a time window that does not exceed the set of coordinates' arrival period, which is set to 1 min in this paper.

The algorithm is required to tackle a computational complexity problem related to the number of pairs of points to be treated, since we need to calculate the distance for each of these pairs, as shown in Figure 4. The distance from one point to the other in a pair of points allows evaluating if the two points are in proximity, a condition required to subsequently verify the points that are associated in groups. As the number of points belonging to a collected group increases, so does the number of comparisons necessary to identify these grouped points.

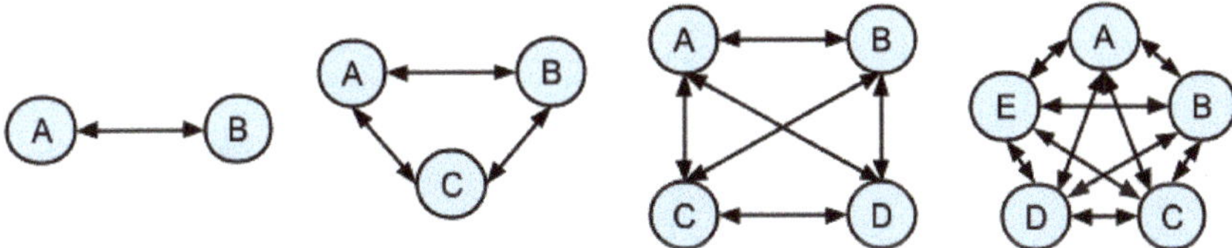

Figure 4. Increase in computation due to the increased amount of points. Source: the authors.

The number of comparisons for the verification of proximity is given by the simple combination formula (Equation (2)) where n is the number of items in a collection and p is the number of elements in each combination, so the result $C_{n,p}$ refers to the combination of n things taken k at a time without repetition. In our calculations, p is set to two since we refer to pairs of points in a sample of coordinates that must be treated each time.

$$C_{n,p} = \frac{n!}{p!(n-p)!} = \frac{n.(n-1).(n-2)!}{2.(n-2)} = \frac{n.(n-1)}{2} \tag{2}$$

For example, in a sample of 10,000 points, there would be approximately 50 million distance calculations. This number of computations and the required processing time window are critical factors for a successful implementation. Furthermore, it is important to control these factors since the amount of monitored individuals can grow with the evolution of an anklet-based monitoring system utilization, given the prison population growth rate, as shown in Figure 1.

The problem is solved partly by dividing the total coordinate space into subareas, which allows breaking down the processing instances, as described in Section 3.1.5. The proposed solution can be completed with the cooperation of parallel computing nodes. Indeed, to prevent the amount of monitored individuals from compromising the processing within a defined time window, the alternative proposal is an algorithm that processes subareas of the coordinate space in parallel. In this case, as the number of points to process grows, one can add more parallel nodes to the system for the completion of the processing within a required time window.

3.1.5. Division of the Coordinate Space into Subareas to Allow Processing Parallelization

The problem of executing a number of proximity-related computations within a required processing time window demands a solution where more computational power can be added to the system when the number of monitored individuals increases or when there is a reduction in the processing time window. Thus, the division of the coordinate area into smaller areas is proposed here so that the processing can be divided into several processing units.

Referring to Figure 5, we consider an initial area computed from the farthest points in a periodical sample reported by the monitored devices. This constitutes the abstraction of a square geographical area containing all of the sample points. Then, a recursive division of this area takes place guided by a divide-and-conquer strategy as follows, also supported by the work from Ding and Densham [21].

Figure 5. Subdivision of the total area into subareas. Source: the authors.

First, the abstracted area is divided into four smaller areas of equal size (quadrants), and the number of points in each quadrant is counted. If it is observed that a quadrant contains more points than a quadrant population threshold, this quadrant will be further divided so that the recursive quadrant divisions result in a number of quadrants, each one containing a number of points that do not represent a performance processing problem regarding proximity calculations within the limited time window. The quadrant population threshold, i.e., the maximum amount of points a quadrant can have, is arbitrarily fixed in this paper, but as this threshold is bound to the available processing capacity, it should be considered as a variable whose behavior is a matter of future study. Furthermore, a quadrant cannot be subdivided if the length of its size is less than the proximity distance threshold.

This recursive subdivision of the original space is similar to that proposed by Xia et al. [23] using a quadtree structure. However, this study does not consider the hierarchical link between subareas. The central interest is that each of these areas can be processed independently from the other areas, which enables processing parallelization. In another alternative view, the distance calculations occur only inside the quadrants where the points are, thus reducing processing effort. However, although we no longer compare points that are from distant quadrants and thus reduce the number of distance calculations, there will be situations where two points are in proximity in adjacent quadrants, and there is a possible identification failure for that pair. Referring to Ding and Densham [21], we have an alternative to solve this problem, by expanding the area of a newly-created quadrant (Figure 5) by adding to it a margin equivalent to the minimum distance for identifying points in proximity (Figure 6).

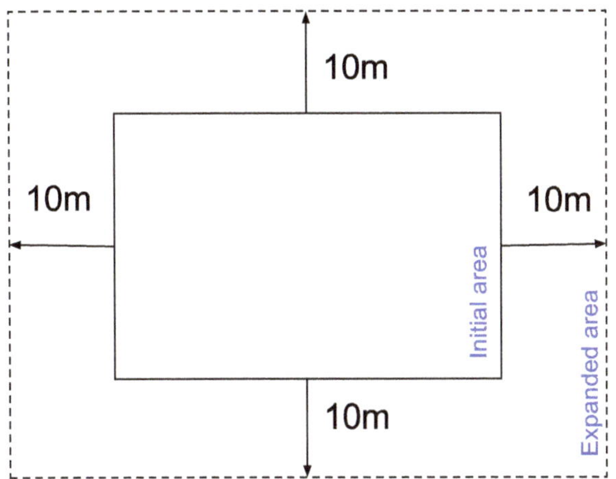

Figure 6. Area expansion to contemplate the proximity of points in adjacent areas. Source: the authors.

Considering this added margin, each area overlaps the adjacent ones, allowing proximity calculations for points that are close to points in adjacent quadrant borders. Since the calculations for a quadrant are independent of those for another quadrant, it is possible to obtain duplicate responses for the same pair of points in proximity (A-B and B-A). Such duplication does not pose a problem as duplicates are eliminated by the groups detection algorithm explained in Section 3.2.4 and shown as Step 3 of the systemic model (Figure 7).

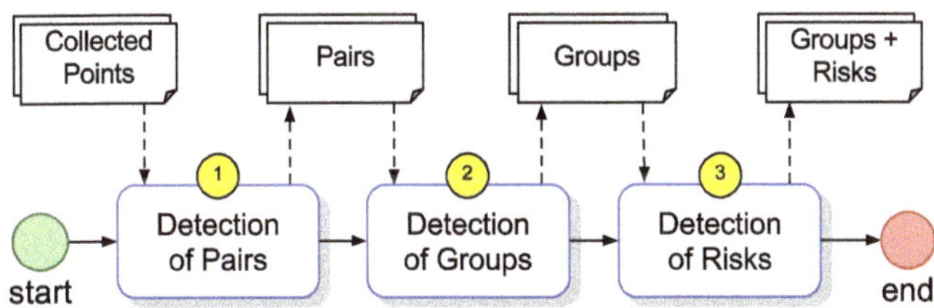

Figure 7. Processing steps. Source: the authors.

3.2. Systemic Model and Associated Algorithms

Figure 7 shows the systemic model proposed in this paper, which divides the processing into three steps: (1) detection of pairs: receive a collection of points sent by the devices through the network, and calculate the points in proximity; (2) detection of groups: group the points in proximity in clusters; and (3) detection of risk indicators: add data on group duration and the number of participants.

The first step receives as input a collection of points collected at a given instant (collected points), whose structure is described in Table 1. The collected points are treated in the second step by the detection of pairs algorithm, which generates a list of points in proximity (Table 1). Subsequently, the detection of groups algorithm examines in Step 2 the list of pairs and generates a list of identified groups whose risk attributes are then calculated in Step 3, resulting in the final output structured as specified in Table 2.

Table 1. Collected points: Algorithm 1 (detection of pairs) input.

Attribute	Type	Description
Device	Integer	Device identifier number
Date/time	Date/time	Date, hour and minute when the point was collected
Point	Geographic coordinate	Geographic point consisting of latitude and longitude

Table 2. Groups and risks: Algorithm 3 (detection of risks) output.

Attribute	Type	Description
Group	Integer	Uniquely identifies the group
Start Date/time	Date/time	Date/time when the group convenes
End date/time	Date/time	Date/time the group dispersed
Processing turn	Integer	Number of processing turns in which the group was detected
Devices quantity	Integer	Cumulated sum of the number of group elements used for calculating the average number of group members
List of devices	List of device identifier numbers	List of devices that have been members of the group

3.2.1. Algorithm 1: Detection of Pairs

Referring to Figure 8, detailing Step 1 of Figure 7, the coordinate points are compared to each other, and the pairs whose distance is less than or equal to lr (Equation (1)) are identified as pairs of points in proximity and added to a list that will be part of the output of the algorithm. In this case, it it worth remembering the distance considered for this approach is 20 m due to precision errors that occur in GPS systems as detailed in [22]. Its input is a list of points collected with the structure detailed in Table 1, and its steps are detailed in Table 3.

This step is the most costly in computational terms, as it implies the comparison between all of the points in the sample to identify points in proximity (Figure 4).

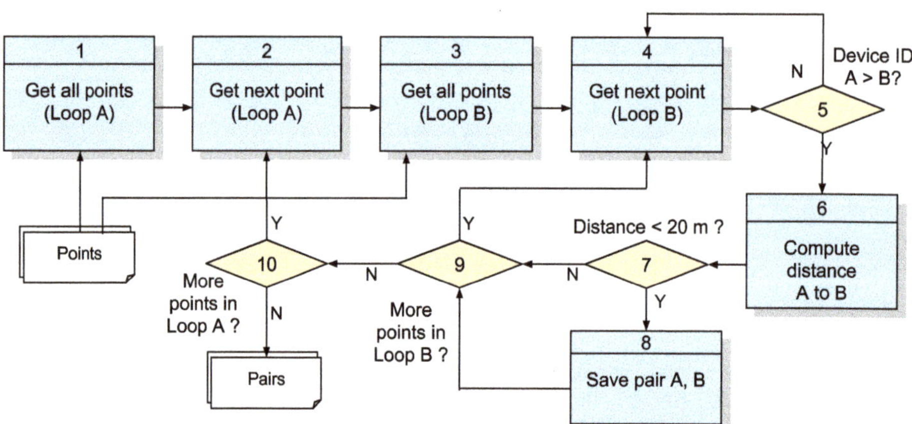

Figure 8. Algorithm 1: detection of pairs. Source: the authors.

Table 3. Detailed steps of Algorithm 1 (detection of pairs).

#	Description
1	Since it is necessary to compare the points to each other, the process starts a loop considering all collected points.
2	It takes each point obtained in the previous item relative to Loop A.
3	It sweeps again all collected points (Loop B) to be compared with Loop A points.
4	It takes each point obtained in the previous item relative to Loop B.
5	This filter prevents the calculation of the distance from A to B to be repeated for B to A. If A is greater than B, it ignores this pair and goes back to Step 4.
6	It uses a geoprocessing function and obtains the distance from A to B in meters.
7	If the distance is greater than lr meters, then Points A and B are not in proximity, and the flow proceeds to the next point to be used in Loop B.
8	If Points A and B are at lr meters or less away from each other, then they are considered to be in proximity and are recorded/stored for the grouping step.
9	If there are more points relative to Loop B to be compared, then it diverts the flow to capture the next Point B.
10	If there are more points relative to Loop A to be compared, then it diverts the flow to capture next Point A. If there are no more points, the comparison processing is completed, and as a result, it outputs records with pairs of devices in proximity.

3.2.2. Algorithm 1.1: Recursive Division of the Original Space into Subareas

Given the concepts presented on the definition of proximity and the idea of dividing the space into smaller quadrants as a function of the number of points to be treated, we have devised the algorithm shown in Figure 9, which is responsible for receiving the collected points, then defining the adequate quadrants and listing the points that are inside these quadrants (Figure 5).

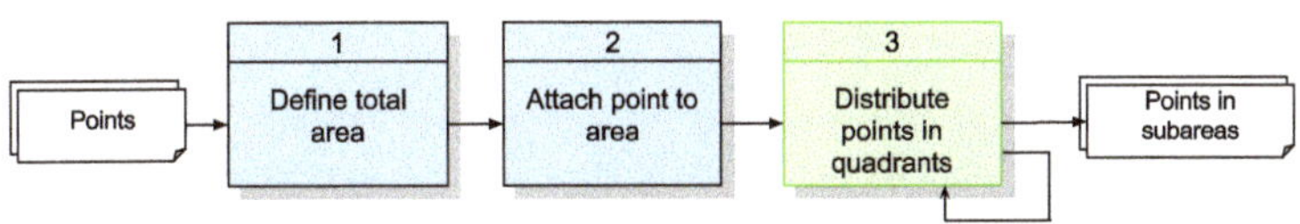

Figure 9. Algorithm 1.1: distribution of points into subareas. Source: the authors.

The first two functions define an initial area covering all collected points and links all points to this initial area. This is necessary so that Function 3 can work recursively. Function 3 always receives an area with its points and then makes a decision whether it is necessary to subdivide this area into smaller quadrants. The decision criterion stipulates that if the number of points inside the area exceeds the quadrant population threshold, this area must be subdivided into smaller areas that will be recursively submitted to Function 3. The details of this algorithm are specified in Table 4.

Table 4. Detailed steps of Algorithm 1.1: division into subareas.

#	Description
1	It obtains the most distant points of the map and generates a square geographical area that covers all points to be processed.
2	It links the points to be processed to the area created in the previous item. This step is required as a preparation for the first call to the recursive function described in the next item. The input parameter for this function is the area with its collection of points.
3	The recursive function divides the received area into quadrants (four new areas) and modifies the links of the points from the received area to the new quadrants according to the coordinates of these points.

The recursive function shown in Figure 9 has its algorithm shown in Figure 10, while Table 5 describes the steps of this recursive function.

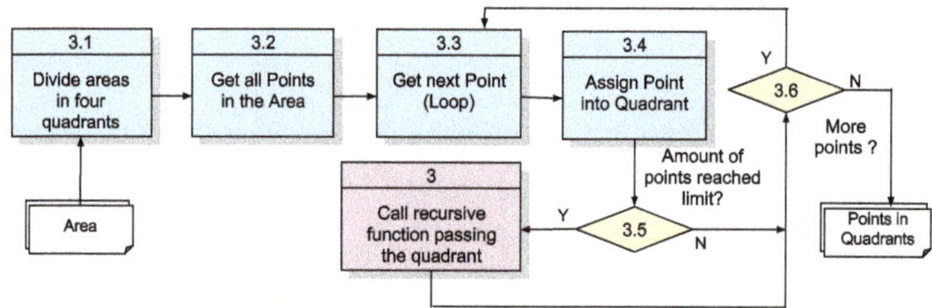

Figure 10. Flow of the recursive space division function in Algorithm 1.1. Source: the authors.

Table 5. Steps of the recursive space division function in Algorithm 1.1.

#	Description
3.1	The area received as input is divided into four quadrants. Recursively, this division is performed for the initial processing of the total area or when one of the quadrants has a number of points that exceeds the quadrant population threshold. Each created quadrant has its area expanded to contemplate proximity among points that are close to each other, but pertain to different adjacent quadrants.
3.2	It reads all of the points that are linked to the area provided as input, preparing then to move each of the new quadrants according to their coordinates.
3.3	With the points supplied in the previous item, a finite loop is run to assign the appropriate treatment.
3.4	According to the coordinate of the point, it is copied from the input area to the corresponding quadrants. Due to the area expansion of each quadrant performed in 3.1, a point may appear in more than one quadrant.
3.5	During the loop, if the number of points linked to a quadrant reaches a maximum value, this quadrant is used as input for a recursive call to further divide this quadrant.
3.6	If there are still points to be treated, the loop is repeated for the next point.

Each subarea set of points can be assigned to be processed on different computational nodes, which allows the work to be parallelized. While on the one hand, we ensure that each subarea has a number of points smaller than an established threshold, on the other hand, we may have subareas with a small number of points. This may represent a potential waste of computing and memory resources since the processing varies according to the number of points in the subareas. However, a scheduling process was adopted in this work that distributes sequentially the subareas in the available threads, minimizing possible differences in the total processing time in the nodes.

Algorithm 1.1 generates a list of points with their respective subareas to be processed in parallel by Algorithm 1.2. Such a structure is detailed in Table 6.

Table 6. Groups and risks: Algorithm 3 (detection of risks) output.

Attribute	Type	Description
Device	Integer	Device identifier number
Date/time	Date/time	Date, hour and minute of point collection
Point	Geographic coordinate	Geographic point consisting of latitude and longitude
Subarea	Integer	Identifier of the subarea containing the point

3.2.3. Algorithm 1.2: Detection of Pairs within Subareas

The algorithm for the detection of pairs within subareas is the same described in Figure 8, only differing on the list of points to be processed. The output of this algorithm is a list containing pairs of points in proximity whose structure is detailed in Table 7.

Table 7. Algorithm 1.2 output structure (detection of proximity pairs).

Attribute	Type	Description
Pair	Integer	Pair identifier number
Date/time	Date/time	Date, hour and minute of point collection
Device A	Integer	Device identifier number
Point A	Geographic coordinate	
Device B	Integer	Device identifier number
Point B	Geographic coordinate	Geographic latitude and longitude of the point

3.2.4. Algorithm 2: Detection of Groups

This algorithm, Number 2 of Figure 7, takes the pairs of points considered in proximity by Algorithm 1.2 and then finds those that are grouped by looking for neighbors of a neighbor, i.e., in situations where Point A is close to B and Point B is close to C; hence, A, B and C form a group of monitored individuals. This detection of groups algorithm is presented in Figure 11, while its details are specified in Table 8 and its output in Table 9 with a list of groups, each one having an identifier, a timestamp for the moment the points were collected and a list of devices composing the group.

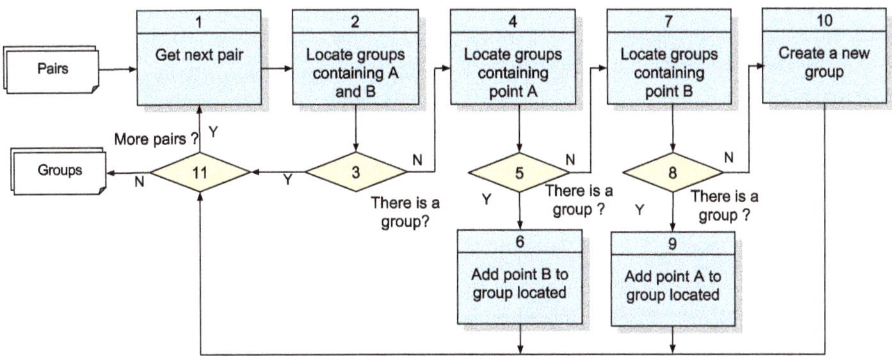

Figure 11. Algorithm 2: detection of groups. Source: the authors.

Table 8. Detailed steps of Algorithm 1 (detection of pairs).

#	Description
1	It obtains each pair, A and B points, of the set obtained in the previous step.
2	It checks for any group from previous iterations that already has Points A and B and eliminates any possible repetition.
3	If a group with the 2 points is identified, then nothing needs to be done, and the loop must continue to the next pair.
4	It checks for any group that has at least Point A.
5	If the group is located, then it does not have Point B.
6	It adds Point B to the group located in the previous item.
7	It checks for any group that has at least Point B.
8	If the group is located, then it does not have Point A.
9	It adds Point A to the group located in the previous item.
10	If no group is located containing either A or B, then a new group must be created with the A and B pair. This group can then be completed as new points are discovered in later iterations.
11	If there is still a pair to be processed, then it processes the next pair; else the algorithm ends.

Table 9. Algorithm 2 output structure (detection of groups).

Attribute	Type	Description
Group	Integer	Group identifier number
Date/time	Date/time	Date, hour and minute of point collection
Device list	Integer list	List of devices that make up the group

3.2.5. Algorithm 3: Computation of Risk Indicators

Detection of risks, which corresponds to Algorithm 3 in Figure 7, computes for a group of monitored individuals additional data regarding the duration of the group and the average number of elements, indicators that are updated as new samples are collected from monitored devices. From the standpoint of anklet monitoring, these data about groups may contribute to the risk analysis to be conducted subsequently to performing the specified algorithms. The proposed solution in this paper just computes the risk indicators linked to identified groups and stores these data for a risk analysis activity to be performed outside the monitoring system.

During its execution (Figure 12 and Table 10), Algorithm 3 collects the following data: (i) group duration: this indicator comes from the perception that groups that last longer may be indicative of greater risk and even that groups with a very short duration may be discarded; (ii) the average number of elements in each group: groups with a higher number of elements can indicate larger scale violations involving, for example organized crime or conspiracy.

In order to indicate the duration or permanence of a group, the algorithm must update this previously identified group with data regarding duration (start/end date/time). When the end date/time attribute is not populated, it indicates that the group is still active, i.e., it has been continuously sustained until the last data fusion execution. Registered dates/hours for the group do not represent the exact instant of this group start or end, as they are influenced by wait and service times during sensor data collection and the execution of the algorithms themselves.

As discussed before, the output of Algorithm 3 is specified in Table 2. The resulting structure is then available for risk analysis and for future processing turns.

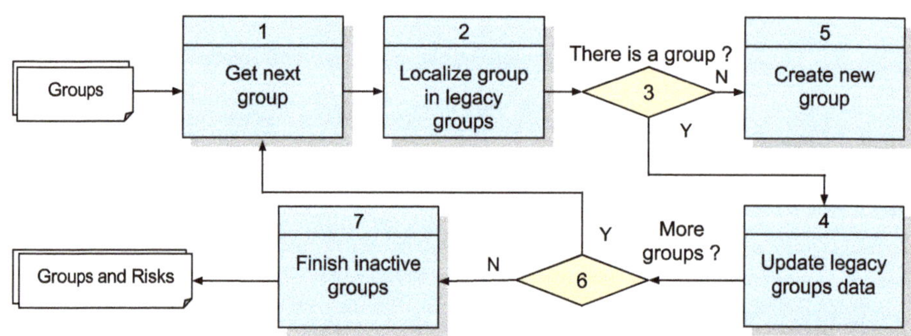

Figure 12. Algorithm 3: computation of risk indicators. Source: the authors.

Table 10. Algorithm 3 specification (group risk indicators).

#	Description
1	It obtains each group detected in the current processing turn.
2	In the set of previous active groups, it identifies the groups that possess at least 50% of its members in common with a group detected in this processing turn. The 50% percentage parameter is an arbitrary choice to be further investigated in future studies.
3	If no compatible group is found, then the flow is directed for the creation of a new group in Item 5.
4	It updates the number of processing turns for the group and adds up the number of elements counted in each processing turn. These two attributes provide the necessary data for averaging the number of members of the group during its existence.
5	It creates a new group considering the processing count attribute as 1, the start date/time attribute as the timestamp of the current processing and the elements count attribute as the respective number of group members.
6	If there are more groups to be processed, then it directs the flow to capture the next group; else it then closes the existing groups, as per Step 7.
7	Existing groups that were not identified in the current processing turn should be ended. This is done by updating the group's end date/time with the value corresponding to the immediately previous processing turn, i.e., the last time the group was detected. As a result of the processing, a set of groups is generated, as well as their duration and number of members for utilization in the next processing turn.

4. Validation Scenarios and Results

In order to validate the algorithms presented in this paper, a simulated database was used with approximately 10,000 devices. The simulation of groups was performed by creating variations of a set of paths obtained from real GPS equipment. The simulated new routes were composed using the horizontal and vertical displacement of the original device coordinate points in the geographic space, also increasing the number of coordinate points in the sample. Moreover, new routes were created by reversing the latitudes and longitudes and attributing them to new simulated points. As a result, three sets with 10,000 points each were generated. These samples correspond to three consecutive collections of points from simulated anklets in a simulated schedule, respectively corresponding to the date 25 May 2015 at time tags: (i) 12:00, (ii) 12:01 and (iii) 12:02. In Table 11, there is a sampling of records randomly extracted from the simulated database.

Table 11. Sampling extracted from simulated database.

Timestamp	Device ID	Latitude	Longitude
25 May 2015 12:00:00–03	12133	−29.903980255127	−51.169883728027
25 May 2015 12:00:00–03	1096	−30.062665939331	−51.192127227783
25 May 2015 12:00:00–03	41978	−29.778089523315	−51.108917236328
25 May 2015 12:00:00–03	817	−30.028823852539	−51.225776672363
25 May 2015 12:00:00–03	40413	−30.093103408813	−51.177989959717
...	...	...	...
25 May 2015 12:01:00–03	12123	−30.087636947632	−51.231784820557
25 May 2015 12:01:00–03	10871	−30.049358333333	−51.162086666667
25 May 2015 12:01:00–03	1969	−30.114995956421	−51.362251281738
25 May 2015 12:01:00–03	91523	−29.70588684082	−53.802436828613
25 May 2015 12:01:00–03	91042	−30.050704956055	−51.21089553833
...	...	...	...
25 May 2015 12:02:00–03	5575	−30.016288757324	−51.11653137207
25 May 2015 12:02:00–03	11716	−30.062965393066	−51.142623901367
25 May 2015 12:02:00–03	12165	−30.201919555664	−51.134094238281
25 May 2015 12:02:00–03	5954	−29.986715316772	−51.1682472229
25 May 2015 12:02:00–03	1047	−30.086135864258	−51.234657287598
...	...	...	...

As an example of a coverage area, the 10,000 points regarding Timestamp 25 May 2015 12:00:00-03 sampling are spread in a geographic area as illustrated in Figure 13.

Figure 13. Sampling points at 25 May 2015 12:00:00-03. Source: the authors.

The computing configuration used to validate the proposed algorithms execution is presented in Table 12.

Table 12. Equipment used in tests.

Resource	Specification
Processor	Core i5-2467M 1.6 GHz (dual core with hyper-threading)
Memory	8 Gigabytes
Hard drive	516 Gigabytes 5400 RPM hard disk
Operating system	Ubuntu 14.04.4
DBMS	PostgreSQL 9.4

The proposed algorithms are implemented and evaluated in five scenarios. The first one is a serial implementation in a relational database query language (scenario). This classical scenario for application development is taken as a baseline for comparing the results in this paper since it does not deploy any particular performance contribution, though it presents the complete correct functionality proposed in this paper. The other scenarios gradually present the contributions of parallelism (Scenario 2) and programming language (Scenarios 3 and 4) for the proposed algorithms, maintaining the same functionality. The possible distributed processing scenario is analyzed in the discussion of the results.

Specifically, for correction purposes, in all evaluated scenarios, the number of records resulting from the execution of each algorithm applied to the simulated data is given in Table 13.

Table 13. Results of the algorithms' processing.

Algorithms	Collection of Points		
	12:00	**12:01**	**12:02**
Number of Subareas	22	37	91
Detection of Pairs	9103	9580	14,686
Detection of Groups	1673	1762	1854
Detection of Risks Indicators	406 active groups in all collections		

In the presentation of each scenario's results hereafter, measurement values are the average of 10 repeated executions.

4.1. Scenario 1: Serial Processing in PL/pgSQL

In this scenario, the described Algorithms 1–3 are fully executed in serial processing, and the performance of each one is measured. Each of the three algorithms is measured separately, and then, their summed response time is presented. This method is chosen to enable reasonable comparisons with results from subsequent scenarios, when parts of those algorithms are replaced by modules in parallel processing or in C language or in distributed processing.

4.1.1. Algorithm 1: Detection of Pairs

The implementation is of a simple algorithm that compares all of the coordinate points by obtaining a list of pairs of points in proximity generated via a SQL statement that performs a self-join on the table of points. In this SQL command, a filter in the where clause selects only the points whose calculated distance is less than the proximity threshold. This threshold was defined at 20 m as stated in Section 3.1.1. Furthermore, in this where clause, a filter is added that considers only the points where the ID of a point A is smaller than the ID of a point B. This filter prevents calculating two times the distances between the same pair, i.e., distance from A to B and from B to A, thus reducing processing effort. As output, a table of pairs is generated. For instance, the output table, corresponding to our sample tagged 25 May 2015 12:00, has an approximate number of 9103 records (pairs of points in proximity).

This algorithm was tested with distance varying 10, 20 and 40 m as the threshold. Though there was variation on the amount of pairs detected because of the distance variable, the response time remained approximate. Besides, 20 m is acceptable within the error precision [22].

4.1.2. Algorithm 2: Detection of Groups

In order to identify groups of points in proximity, the developed the PL/pgSQL code specified in Figure 11 generates a table of groups with 1673 identified groups.

4.1.3. Algorithm 3: Computation of Risk Indicators

This algorithm is a PL/pgSQL module according to Figure 12. However, the implementation language allows a code improvement by applying an update on the set of records that meet the filter instead of checking each group obtained in the recent processing against each of the previously detected groups.

4.1.4. Response Time Results

Notably, the algorithm that detects pairs of points in proximity presents a much higher processing time than Algorithms 2 and 3, reaching an average time of 280 s (Figure 14). When considering the required overall performance threshold (fixed to a 1-min processing window), the sum of times from the three processing algorithms exceeds this value, which forewarns of their impracticality. However, the measured values are interesting as a baseline for the subsequent validation scenarios.

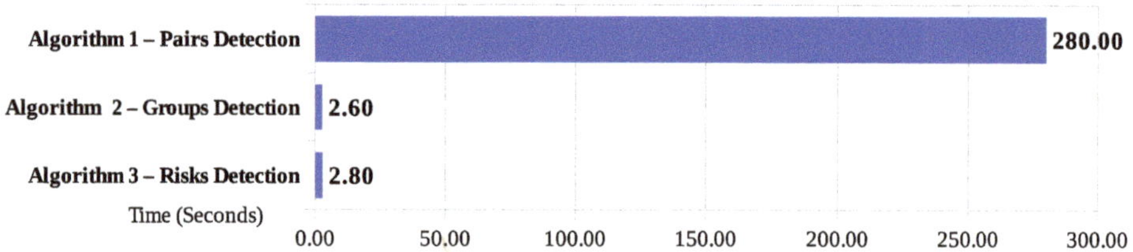

Figure 14. Response time by the algorithms. Source: the authors.

Scenario 1's results illustrate the response time issue when identifying coordinate points in proximity without the use of parallel processing, which justifies the next scenario.

4.2. Scenario 2: PL/pgSQL Processing with Multiple Parallel Instances

In Scenario 1, the Algorithm 1 for the detection of pairs, which takes much more time than the other two algorithms, is the observable candidate for improvement, thus being reformulated in Scenario 2 by adding an inner algorithm to distribute points into subareas (Algorithm 1.1), which allows the identification of pairs (Algorithm 1.2) to be executed in multiple parallel instances. As Algorithms 2 and 3 are not modified from Scenario 1, they are not presented in Scenario 2.

The division of the whole coordinate space into smaller quadrants implies the corresponding division of the number of coordinate points to be compared in each quadrant processing. Now, there is a trade-off regarding the number of points that is used as a decision criterion for recursive sub-divisions of quadrants. It is necessary to set the maximum amount of points per quadrant subarea, considering that the smaller this number, the greater the number of subareas.

Given that we have established a database of 10,000 coordinate points for all validation scenarios, we define four cases for the maximum number of points per subareas (respectively 250, 500, 1000 and 2000) and obtain response time figures for these cases.

4.2.1. Algorithm 1.1: Distribution of Points per Subareas

This algorithm, implemented in PL/pgSQL according to the flows in Figures 9 and 10, based on 10 repeated executions, presents the average response time results shown in Figure 15, for each of the maximum values of points per subarea.

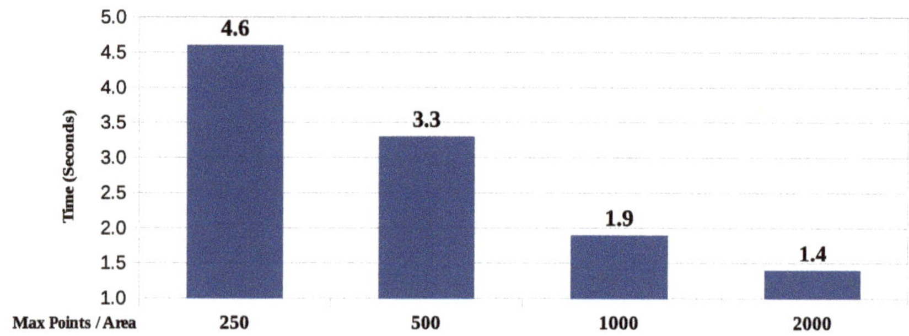

Figure 15. Response time by maximum points by area. Source: the authors.

Data concerning the sub-divisions of the coordinate area are presented in Table 14. As expected, the larger the maximum number of points per area, the less is the average area size to be processed. The size of the largest area is always the same since it corresponds to the first subdivision of the total area corresponding to 146.56 km^2.

The lesser the maximum amount of points per subarea, the smaller the average size of the subareas. The size of the smallest area resulting from the most recursive division into quadrants also decreases with the number of points per area. In the smallest of the cases, the resulting subarea is approximately 150 square meters wide.

Table 14. Maximum points per area and area size.

	Max Points per Area			
	2000	**1000**	**500**	**250**
Number of Subareas	22	37	91	151
Average Area Size (m^2)	6,662,905.40	3,961,789.13	1,610,986.67	970,874.14
Smallest Area Size (m^2)	9023.61	2275.65	578.85	149.74
Largest Area Size (m^2)	36,643,506.10	36,643,506.10	36,643,506.10	36,643,506.10

4.2.2. Algorithm 1.2: Detection of Pairs with Multiple Parallel Instances

The pair detection algorithm from Scenario 1 is adapted to run in parallel. Since PostgreSQL does not support developing routines in PL/pgSQL, we use a shell script running under an Ubuntu operating system that concurrently submits different instances of the same routine so that each instance considers a group of distinct areas.

4.2.3. Response Time Results

Figure 16 shown the compared response time results for the parallel processing taking into account the four values for the maximum points per coordinate subarea (250, 500, 1000 and 2000) and the number of processing threads used (2, 4, 8 and 16).

The amount of subareas will typically be greater than the number of computing nodes (or cores) available to handle them. Despite knowing that it is not the best technique due to variation in the amount of points per subarea (zero to the max points per area), we assume in this paper that the distribution of subareas among the computing nodes will be applied in stages, similarly to a round-robin algorithm. This can result in an overload of specific nodes while others become idle. A better distribution processing technique between nodes is a subject for further work.

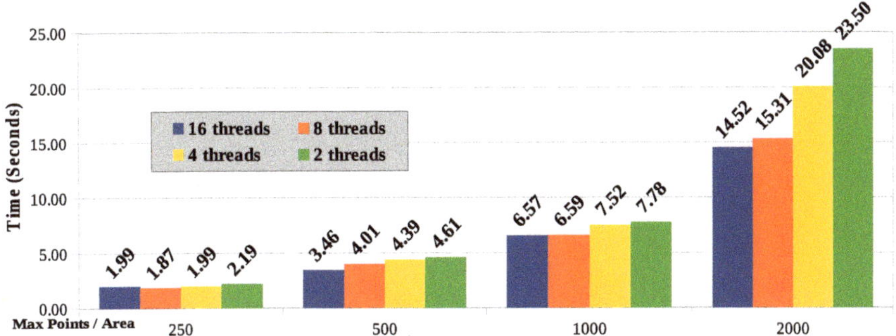

Figure 16. Response times by thread using PL/pgSQL Language. Source: the authors.

Figure 16 shows that as the maximum number of points per area is reduced, so is reduced the corresponding processing time due to the lower number of comparisons between points necessary for processing and calculating the distance. For the sample used in this work and considering the results in Scenario 1, the response time reduction is significant and tends to flatten with the reduction of points per subarea.

The equipment used has processors with two cores and hyper-threading technology that simulates four logical cores. It was expected that the best response time would be with four threads. However, for areas with a higher number of points (1000 and 2000), the cases with eight and 16 threads showed better results. For areas defined with less points, the difference regarding the number of threads is smaller. It should be taken into account that, as the processing performs some disk read and write operations, the consequent I/O wait time seems to explain the better response time when processing with more threads.

4.3. Scenario 3: Algorithm 1.2 in C Language without Parallel Processing

In this scenario, for comparison, a routine was developed using the C language for implementing the algorithm for the detection of pairs (Figure 9) without parallel processing. All points are compared with the others identifying those who are in proximity by calculating the respective distance. In this case, there is no division of points into subareas, and the entire process is performed serially. The routine reads the 10,000 points from a file in a file system and writes the result as a text file in the same file system.

The processing response time (Figure 17) for Algorithm 1.2 in C language was approximately 20-times faster than the same routine in PL/pgSQL (14.66 s in Scenario 3, while it was 280 s in Scenario 1). Of course, this superiority is expected in terms of performance, thus indicating which language is most appropriate for this class of application.

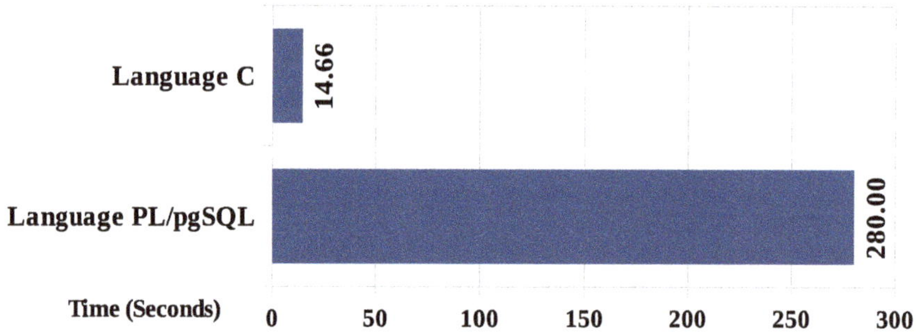

Figure 17. Response time by language.

4.4. Scenario 4: Algorithm 1.2 in C Language with Multiple Parallel Instances

With the routine in C language developed in the previous scenario being executed now in multiple instances separated by coordinate subarea and running such instances with 2, 4, 8 and 16 threads, the response time results are shown in Figure 18.

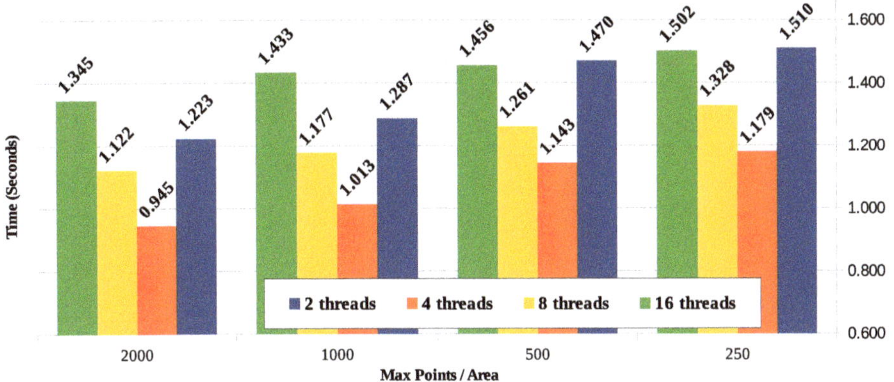

Figure 18. Pair detection algorithm response time in C language.

The results obtained for the parallel execution of the routine in C language, in every case under 2 s, are generally well below those obtained in PL/pgSQL. A noticeable observation is that, unlike the results obtained in processing with PL/pgSQL, the differences are significant regarding the number of running threads, with the best performance being achieved when running with four threads. The lower response time obtained with four threads is justified because of the architecture of the processor of the equipment used in the test, which has four logical cores with hyper-threading technology (two physical cores).

4.5. Discussion

The proposed algorithms were applied on the three sets of points for checking their results and response time in the different scenarios. In the first scenario, we address the application of algorithms directly on the PostgreSQL database using PL/pgSQL and the PostGIS extension.

In the second scenario, the processing was divided into two steps seeking to reduce the overall run time of the algorithms. In this scenario, the proposed solution is the distribution of coordinate points into subareas of the original area allowing processing parallelization for these subareas.

In the third scenario, without parallel processing, the slower task was implemented in C language, but it was found that even in a higher performing language, response time could still be enhanced.

Consequently, in the fourth scenario, the pair identification algorithm implemented in C language was run in multiple parallel instances, giving way to better response time results.

When comparing the lowest possible processing times for each scenario (including the set of algorithms for the identification of pairs of points in proximity, grouping of these pairs and identification of risk indicators), we obtain the graph shown in Figure 19. Scenarios 1 and 3, corresponding to processing without parallelism in PG/pgSQL and in C, respectively, have a final result with higher response times. Scenarios 2 and 4 showed better response times, which became possible due to the utilization of parallel processing for the identification of proximity among coordinate pairs.

In all scenarios, the processing of Algorithms 2 and 3 (detection of groups and computation of risk indicators) is performed with PL/pgSQL language modules due to the good response time obtained with this programming language, which allows the minimum total time to stay at 7.75 s (Scenario 4). Future implementation with all algorithms in C language would further reduce the shortest execution time of the whole process.

This paper does not address the integration of routines in C language being activated by calls from PL/pgSQL functions, a feature supported by PostgreSQL 9. It is estimated, however, that the execution times of the routines in this situation are very close to those measured in our presented scenarios.

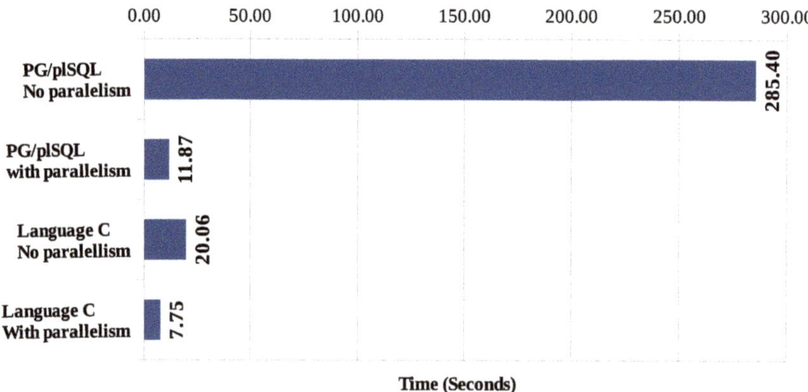

Figure 19. Lowest processing time by scenario.

4.6. On the Feasibility of Using Distributed Processing

As parallel processing with multiple threads performed better compared to single thread, both in PL/pgSQL and in C language, it is therefore natural to think of an experiment in a distributed processing environment in a big data-oriented architecture. In this context, one of the most frequently-used platform is Hadoop. However, some features of this type of processing should be considered:

(a) Big data assumes a massive amount of data to be processed. It seems that this is not the case described in this paper. Although high performant processing is implied by application requirements, the amount of data processed at a time (for instance, the 10,000 points proposed here) is not an impressive data volume. As a result, without any special configuration, loading these data into a multiple node environment as expected in a big data environment, this volume of data tends to be loaded on a single node, thus eliminating the possibility of distributed processing. As discussed by Davenport [24], the term big data is basically defined in terms of volume, variety and velocity, characteristics that guide the implementation of big data platforms. The first characteristic (volume) is already compromised in our use case.

(b) To be treated in a distributed processing environment, the whole process takes a few seconds (in some cases, even minutes) to be ready for processing. This necessary initial time can make it unfeasible to meet the initial requirement described in our problem, which assumes a 1-min time window to perform data samples' processing, in our case with the 10,000 simulated coordinate points for our validation.

It seems that the applicability of the presented algorithms for distributed processing with the big data Hadoop platform, though it can be useful if anklet monitoring is expanded to larger populations, is still an issue to be investigated, since although the offered processing capacity is relatively higher, the amount of data to be processed is low, but requires important preparation effort, which makes big data for now unsuitable as an alternative solution to the problem.

5. Conclusions

The process of monitoring convicts by means of electronic anklets can be improved by producing additional data to support risk analysis for decision making. In this paper, the challenge of identifying a gathering of groups of monitored convicts, the time of permanence of these groups and the number of their members was proposed with the addition of a limited processing time restriction. The use of serial algorithms was shown to be a problem due to their exceeding processing response time. It was observed that the longest processing time concerned the calculation of pairs of device coordinates regarding their proximity.

The proposed solution to increase performance is the division of the total geographical area containing all coordinate points into smaller areas (quadrants) so that each area can be processed independently, thus allowing parallel processing when identifying points of proximity. Dividing the total area into smaller areas involved dealing with the situation where the points in proximity were in adjacent subareas, which was solved by expanding each area by the GPS precision factor (10 m) in all directions and the elimination of duplicates in the grouping of points in proximity.

The adoption of routines using PL/pgSQL for implementing the algorithms alone would not meet the required time window. However, when using a low-level language, such as C, to implement the same algorithms, the overall response time experiences a substantial reduction.

This response time reduction, however, does not justify giving up parallelism in the proposed processing, since even the routine time in C language without using parallelism (which corresponds to approximately 1/4 of the defined window limit) could compromise the performance requirement. For instance, a linear increase in the number of points to be processed increases exponentially the number of comparisons to be performed to calculate the distance, which directly impacts response time. In this case, even adopting the C language to implement the routines relating to the proposed algorithms, it is appropriate to use a parallel processing solution.

Computing with Graphics Processing Units (GPU) is appropriate in cases of short and parallel routines, as is the case of the detection of nearby points by calculating and comparing distances. Although restricted to specific hardware, but abundantly available on the market, this alternative should be considered in the case of the need for even greater reduction in the response time for the algorithms addressed in this paper. While GPU-accelerated computing should be considered in future works due to the intense and parallel processing characteristics or the pair detection algorithm, other big data platforms, such as Hadoop, can be further studied and tested, in order to address a simplified manner to reach the process performance requirements.

Unlike other algorithms, the solution proposed in this paper includes the monitoring of formed groups over time, periodically updating the data of each group, thus supporting the analysis based on group duration (the time interval in which the group remains assembled) and on the average number of elements of the group during its existence. Moreover, when considering inactive groups (those that have been identified in the past and are now ended), the frequency and time at which certain groups usually meet can also be informed.

We emphasize that this paper is dedicated to issues related to proximity calculations and their performance, although we recognize that there are several other issues equally important to be treated, such as the date-related fusion aspects addressed by Khalegui [9] who classified the related questions as imperfection, correlation, inconsistency and disparateness issues. Thus, the data used in our simulations could be modified and completed to represent situations prone to these problems. Then, for our algorithms to be tolerant to these data quality problems, they must include filters with respect to data characteristics, as for instance invalid dates, eventually missing points, repeated or delayed point collections, etc.

Other issues related to the mobility of groups will be addressed in future work due to the complexity of their identification and treatment. Such problems occur, for example, when there is a tracked device in a meeting and this device fails to submit the coordinates, thus causing the wearer to be considered outside the group in the corresponding periodic processing. Indeed, groups identified and tracked, being stationary or on the move, are handled the same way in the proposed algorithms, although each of these situations may pose different risks.

The integration of the data produced in this work with other complementary databases, such as those registering crimes occurring at the same time and location of the meetings detected, and the registry of individual dangerousness are important to increase and improve the information made available to the investigation teams. Such integration should be addressed as future work.

Acknowledgments: This research work has the support of the Brazilian research and innovation Agencies CAPES - Coordination for the Improvement of Higher Education Personnel (Grant 23038.007604/2014-69 FORTE - Tempestive Forensics Project) and FINEP - Funding Authority for Studies and Projects (Grant 01.12.0555.00 RENASIC/PROTO - Secure Protocols Laboratory of the National Information Security and Cryptography Network), as well as the Brazilian Ministry of Justice (Grant 001/2015 SENACON - National Consumer Secretariat), the Ministry of Planning, Development and Management (Grants 005/2016 DIPLA - Planning and Management Directorate, and 11/2016 SEST - State-owned Federal Companies Secretariat) and the DPGU - Brazilian Union Public Defender (Grant 066/2016). This work was also funded by the European Commission Horizon 2020 Programme under Grant Agreement Number H2020-FCT-2015/700326-RAMSES (Internet Forensic Platform for Tracking the Money Flow of Financially Motivated Malware).

Author Contributions: Paulo Lima Machado, Rafael T. de Sousa Jr., Robson de Oliveira Albuquerque and Luis Javier García Villalba are the authors who mainly contributed to this research, performing experiments, analysis of the data and writing the manuscript. Tai-Hoon Kim analyzed the data and interpreted the results. All authors read and approved the final manuscript.

Conflicts of Interest: The authors declare no conflict of interest.

References

1. DEPEN, Levantamento Nacional de Informações Penitenciárias. Available online: http://dados.mj.gov.br/dataset/infopen-levantamento-nacional-de-informacoes-penitenciarias (acessed on 27 March 2016).

2. Barbosa, R.M. *O Monitoramento Eletrônico Para Presos de Baixa Periculosidade*; Universidade Católica de Brasília: Brasília, Brazil, 2010. (In Portuguese)

3. Daubal, M.; Fajinmi, O.; Jangaard, L.; Simonson, N.; Yasutake, B.; Newell, J.; Ali, M. Safe step: A real-time GPS tracking and analysis system for criminal activities using ankle bracelets. In Proceedings of the 21st ACM SIGSPATIAL International Conference on Advances in Geographic Information Systems, Orlando, FL, USA, 5–8 November 2013; p. 515.

4. Federación Española de Empresarios de Automocion. *Congreso Nacional de Execução de Penas e Medidas Alternativas*; Federación Española de Empresarios de Automocion: Manaus, Brazil, 2008. (In Portuguese)

5. Vani, B.; Monico, J.F.G.; Shimabukuro, M.H. *Fundamentos e Aspectos Computacionais Para Posicionamento Por Ponto GP*; Revista Brasileira de Geomática: Apucarana, Brazil, 2013. (In Portuguese)

6. Nathan, P.M.C. *Wireless Communications*; PHI Learning Pvt. Ltd.: Delhi, India, 2010.

7. Moss, M.L; Townsend, A.M. How telecommunications systems are transforming urban spaces. In *Cities in the Telecommunications Age: the Fracturing of Geographies*; Routledge: New York, NY, USA, 2000.

8. Hall, D.L.; Llinas, J. An introduction to multisensor data fusion. *Proc. IEEE* **1997**, *81*, 6–23.

9. Khaleghi, B.; Khamis, A.; Karray, F.O.; Razavi, S.N. Multisensor data fusion: A review of the state-of-the-art. *Inf. Fusion* **2013**, *14*, 28–44.

10. Liu, Q.; Deng, M.; Shi, Y.; Wang, J. A density-based spatial clustering algorithm considering both spatial proximity and attribute similarity. *Comput. Geosci.* **2012**, *46*, 296–309.

11. Chew, L.P. Constrained delaunay triangulations. *Algorithmica* **1989**, *4*, 97–108.

12. Carlino, G.A.; Carr, J.K. *Clusters of Knowledge: R&D Proximity and the Spillover Effect*; Business Review; Federal Reserve Bank of Philadelphia: Philadelphia, PA, USA, 2013.

13. Louhichi, S.; Gzara, M.; Ben Abdallah, H. A density based algorithm for discovering clusters with varied density. In Proceedings of the 2014 World Congress on Computer Applications and Information Systems (WCCAIS), Hammamet, Tunisia, 17–19 January 2014.

14. Morreale, P.; Suleski, R. System Design and Analysis of a Web-Based Application for Sensor Network Data Integration and Real-time Presentation. In Proceedings of the 3rd Annual IEEE International Systems Conference, Vancouver, BC, Canada, 23–26 March 2009.

15. Zhu, S.; Kong, L.; Chen, L. Data Mining of Sensor Monitoring Time Series and Knowledge Discovery. In Proceedings of the 2011 2nd International Conference on Artificial Intelligence, Management Science and Electronic Commerce (AIMSEC), Dengfeng, China, 8–10 August 2011.

16. Jakkhupan, W.; Klaypaksee, P. A Web-based Criminal Record System Using Mobile Device: A Case Study of Hat Yai municipality. In Proceedings of the 2014 IEEE Asia Pacific Conference on Wireless and Mobile, Bali, Indonesia, 28–30 August 2014; pp. 243–246.

17. Sathyadevan, S.; Devan, M.S.; Gangadharan, S.S. Crime Analysis and Prediction Using Data Mining. In Proceedings of the 2014 First International Conference on Networks & Soft Computing (ICNSC2014), Guntur, Andhra Pradesh, India, 19–20 August 2014; pp. 406–412.
18. Oduor, C.; Acosta, F.; Makhanu, E. The Adoption of Mobile Technology as a Tool for Situational Crime Prevention in Kenya. In Proceedings of the 2014 IST-Africa Conference Proceedings, Le Meridien Ile Maurice, Pointe Aux Piments, Mauritius, 7–9 May 2014.
19. Park, K.Y.; Youn, H.Y. Crime Prevention System based on Context-Awareness. In Proceedings of the 2011 3rd International Workshop on Intelligent Systems and Applications, Wuhan, China, 28–29 May 2011.
20. Urbano, F.; Dettki, H. Storing tracking data in an advanced database platform. In *Spatial Database for GPS Wildlife Tracking Data*; Springer: Berlin/Heidelberg, Germany, 2014; pp. 9–24.
21. Ding, Y.; Densham, P.J. Spatial strategies for parallel spatial modelling. *Int. J. Geogr. Inf. Syst.* **1996**, *10*, 669–698.
22. NSTB/WAAS T&E Team. *Global Positioning System (GPS) Standard Positioning Service (SPS) Performance Analysis Report*; Federal Aviation Administration (FAA): Washington, DC, USA, 2014.
23. Xia, Y.; Liu, Y.; Ye, Z.; Wu, W.; Zhu, M. Quadtree-based domain decomposition for parallel map-matching on GPS data. In Proceedings of the 2012 15th International IEEE Conference on Intelligent Transportation Systems, Anchorage, AK, USA, 16–19 September 2012; pp. 808–813.
24. Davenport, T. *Big Data at Work: Dispelling the Myths, Uncovering the Opportunities*; Havard Business Review Press: Watertown, MA, USA, 2014.

 International Journal of
Geo-Information

Article

Camera Coverage Estimation Based on Multistage Grid Subdivision

Meizhen Wang [1,2,3], **Xuejun Liu** [1,2,3,*], **Yanan Zhang** [1,2,3] **and Ziran Wang** [4]

1   Key Laboratory of Virtual Geographic Environment, Nanjing Normal University, Ministry of Education, Nanjing 210023, China; wangmeizhen@njnu.edu.cn (M.W.); zhanganan@gmail.com (Y.Z.)
2   State Key Laboratory Cultivation Base of Geographical Environment Evolution (Jiangsu Province), Nanjing 210023, China
3   Jiangsu Center for Collaborative Innovation in Geographical Information Resource Development and Application, Nanjing 210023, China
4   Nanjing Normal University Taizhou College, Taizhou 225300, China; wangzr@outlook.com
*   Correspondence: liuxuejun@njnu.edu.cn; Tel.: +86-137-7668-5731

Academic Editors: Chi-Hua Chen, Kuen-Rong Lo and Wolfgang Kainz
Received: 15 December 2016; Accepted: 31 March 2017; Published: 5 April 2017

Abstract: Visual coverage is one of the most important quality indexes for depicting the usability of an individual camera or camera network. It is the basis for camera network deployment, placement, coverage-enhancement, planning, etc. Precision and efficiency are critical influences on applications, especially those involving several cameras. This paper proposes a new method to efficiently estimate superior camera coverage. First, the geographic area that is covered by the camera and its minimum bounding rectangle (MBR) without considering obstacles is computed using the camera parameters. Second, the MBR is divided into grids using the initial grid size. The status of the four corners of each grid is estimated by a line of sight (LOS) algorithm. If the camera, considering obstacles, covers a corner, the status is represented by 1, otherwise by 0. Consequently, the status of a grid can be represented by a code that is a combination of 0s or 1s. If the code is not homogeneous (not four 0s or four 1s), the grid will be divided into four sub-grids until the sub-grids are divided into a specific maximum level or their codes are homogeneous. Finally, after performing the process above, total camera coverage is estimated according to the size and status of all grids. Experimental results illustrate that the proposed method's accuracy is determined by the method that divided the coverage area into the smallest grids at the maximum level, while its efficacy is closer to the method that divided the coverage area into the initial grids. It considers both efficiency and accuracy. The initial grid size and maximum level are two critical influences on the proposed method, which can be determined by weighing efficiency and accuracy.

Keywords: camera coverage estimation; multistage grid subdivision; line of sight; viewshed analysis; obstacle

1. Introduction

Visual coverage is an essential quantifiable feature of an individual camera and camera network, which perform the most fundamental requirements of any surveillance tasks and computer vision applications. Such diverse applications as camera reconfiguration, optimal camera placement, camera selection, camera calibration, and tracking correspondence are required for capturing coverage information, even though they vary in objectives and constraints. Virtually all camera network applications depend on or can benefit from knowledge about the coverage of individual cameras, the coverage of the network as a whole, and the relationships of cameras in terms of their coverage [1]. Camera coverage is always an essential issue in Visual Sensor Network (VSN), Directional Sensor

Network (DSN), and Wireless Multimedia Sensor Network (WMSN). Visual coverage is also an important issue for (geo-tagged) video data models and retrieval [2], video geospatial analysis [3] and the integration of GIS and video surveillance [4–6].

In the surveillance system, the physical coverage is crucial for spatial analysis, for example to determine whether a suspect or vehicle is exactly covered by a certain camera, to count the number of a certain kind of features covered by camera network, and so on. Consequently, the accurate geometry of the individual cameras and camera network is desperately needed. Moreover, the acceptable speed for coverage estimation is crucial when the number of cameras is large or/and the parameters will be changed frequently—for example optimal camera network deployment, camera network reconfiguration, and so on. Consequently, coverage estimation method considering the trade-off of efficacy and accuracy is desirable.

References are seldom explicit concerning the process of estimating coverage even though almost all applications aim to maximize the overall coverage area sometimes with other constraints, which depend upon the specific application. The coverage problem involves camera parameters, a scene model and task parameters [1]. Because the works themselves are very complex and time-consuming, requiring some approximations when dealing with coverage, the camera model and scene model are often simplified according to the task. The camera model is simplified as a fixed-size sector or quadrilateral. The target fields are often considered as a 2D plane with or without obstacles. A few references investigate algorithms for applications such as coverage optimization considering 3D modeling of the monitored area. In experimental applications, the target area is sampled by regularly arranged grids, so the overall coverage of the target area is represented by the coverage of these grids [7]. It is less time-consuming than methods without sampling, but the result is that simulated experiments with the above assumptions are discordant with the actual applications. These works emphasize efficacy rather than accuracy, and the geometry of the individual cameras and camera network is ignored.

In this paper, we estimate camera coverage considering the trade-off of efficacy and accuracy. We propose a grid subdivision algorithm for estimating camera coverage. The main idea is that the surveillance area is divided gradually into grids of multiple grid sizes, while the coverage area depends on the coverage statuses of grids in different subdivision levels for the following reasons: (1) the camera coverage is not large, which demands a high precision data source; (2) a high precision DEM (Digital Elevation Model) is not always accessible; and (3) the occlusions for line of sight (LOS) from cameras to targets, including buildings, vegetation and other surveyed heights, are often stored in vector features. We assume that the cameras are deployed in 3D geographic space while the surveillance area is a relatively flat ground plane with some occlusions such as buildings, trees and others in vector format. It is more suitable for real-world implementations in most city areas where the ground is seldom rolling.

The remainder of the paper is organized as follows. After a literature review of related work in the next section, the method is described in detail in the third section. Performance of the proposed method is validated through experiments with simulated data and cameras deployed in a real geographic space, and the results are evaluated in the fourth section. Finally, concluding remarks and discussions are presented.

2. Related Works

The researchers in VSN, DSN and WMSN often try to find an efficient algorithm to obtain an optimized configuration scheme for a camera network for different tasks, such as optimal placement [7,8], automated layout [9], coverage-enhancement [10–13], coverage improvement [14], planning optimization [15,16], coverage estimation [17,18], optimal deployment [19–21], camera reconfiguration [22], object coverage [23], scalable target coverage [24], resource-aware coverage [25], etc. Hundreds of cameras are engaged and their parameters frequently changed to estimate coverage in real time, which poses substantial computational challenges. Thus, more emphasis is placed on specific camera coverage models for tasks than the optimization algorithms themselves.

The works mentioned above consider the 3D camera model, but the region of interest is simplified as a 2D plane with/without occlusions and sampled by grid points or control points. Even though this is a feasible way to estimate coverage rate and reduce computing time, it results in inaccurate estimation of the geometric shape of a camera or camera network. Ignorance of the coverage geometry cannot benefit camera network visualization, camera spatial retrieval or later spatial analysis.

Camera coverage can be considered as a particular viewshed analysis because it involves not only geographic data but also the imaging principle of cameras. Viewshed analysis is applied more frequently because of the many potential algorithm parameter changes such as altitude offset of the viewpoint, visible radius, location of viewpoints, effect of vegetation, light refraction, and curvature of the earth. The computational bottleneck poses a significant challenge to current GIS systems [26]. Consequently, the classic viewshed algorithms, such as inter-visibility based on LOS, the Xdraw algorithm, and the reference plane algorithm were improved by a variety of algorithms to speed up calculations [27–31]. Some authors proposed effective parallel viewshed algorithms [26,30,32]. Current research mainly focuses on viewshed analysis in terrain models whose data structure is a DEM or TIN (Triangulated Irregular Network). When combined with a Digital Surface Model (or a Digital Terrain Model), the line of sight method is very effective for surveillance camera placement because it allows introduction of some important characteristics of cameras such as the 3D position of each camera, observation azimuth, field of view, the range of the camera, etc. [33]. However, for most public sources of elevation data, the quality is variable and, in some areas, is very poor (especially in some mountain and desert void areas). This implies that in some situations it is difficult to obtain enough elevation points of the region of interest to build a proper DEM [34]. Occlusions including buildings, vegetation and other surveyed heights are often stored in vector features. Argany et al. [35] stated that besides positional accuracy, semantic accuracy, the completeness of spatial information, and the type of spatial representation of the real world is another important issue that has a significant impact on sensor network optimization. An accurate determination of sensor positions in a raster representation of the space such as in 3D city models is more difficult because visibility could be estimated more accurately in vector data [35].

Overall, in VSN, DSN and WMSN, the researchers designed a camera coverage model to meet the demands of specific optimal tasks. Some of them employed 2D camera models with or without occlusions, and some of them presented 3D camera coverage models considering one or more of FOV (Field of View), resolution, focus, angle and occlusions. The criterion to estimate the camera network is the coverage rate that is determined by the coverage of grid points or control points sampled from the region of interest rather than the physical coverage of cameras. In GIS, the researchers implemented various effective viewshed analysis algorithms. In some works, camera coverage is estimated using an ArcGIS tool [36]. However, the estimated coverage does not exactly conform to the projection principles of camera. An accurate and effective method to estimate camera coverage is desirable to visualize a camera's physical FOV and various optimal applications of a camera network.

3. Camera Coverage Estimation

3.1. Overview of the Method

When the target area is sampled into regularly arranged grids of the same size, the grid size is the most important factor for coverage estimation [35]. If it is undersized, the coverage estimation is of high precision and lower computing efficiency. If it is oversized, the coverage estimation is of low precision and higher computing efficiency because some details are ignored. It is hard to balance the precision and computing efficiency when the target area is sampled into grids of the same size. This paper proposes a method to meet this challenge. The proposed method is shown as Figure 1.

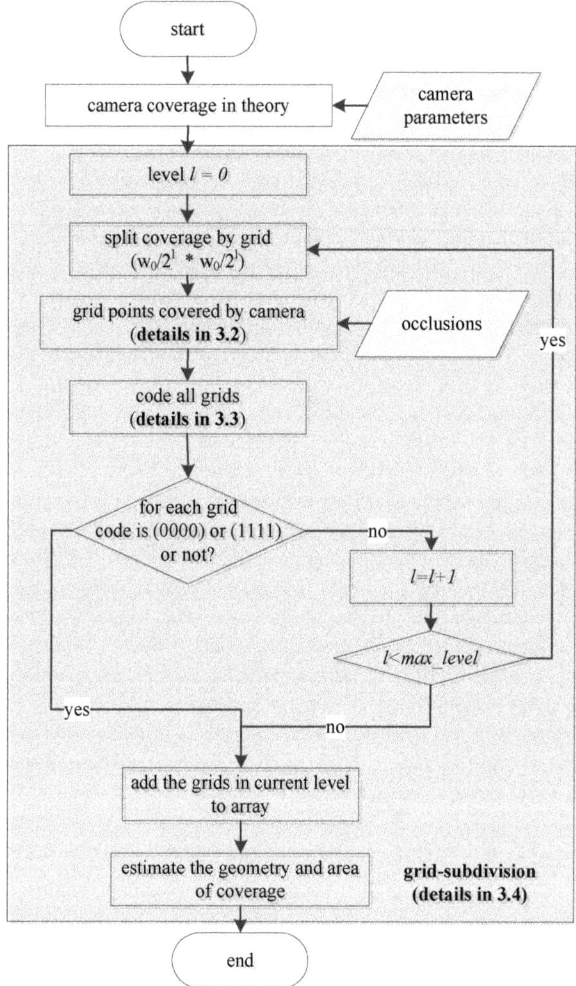

Figure 1. Flowchart of the proposed method.

First, the theoretical camera coverage and its minimum bounding rectangle (MBR) are computed according to camera parameters. Second, the minimum bounding rectangle is subdivided into grids of the initial size written as w_0, and the grid division level, which is written as l, is set to 0. The status of each corner of a grid is estimated by the method depicted in Section 3.2. If a corner point is covered by a camera, then its status is marked as '1'; otherwise, it is marked as '0'. Thus, four digital numbers (0 or 1) are used to code the status of a corresponding grid. Encoding (0000) means that the grid is not covered by a camera and encoding (1111) means that the entire grid is covered. Other encodings such as (0101), (0011), which contain both 0 and 1, mean that the grid is partly covered. The presentation status of a grid is discussed in Section 3.3. Third, each grid in level l whose encoding is not (0000) or (1111) must be subdivided into four sub-grids. The sub-grids will be divided until encoding is (0000) or (1111). Infinite subdivision is not appropriate because it is time-consuming and does not increase accuracy. We stop subdivision when the division level l reaches the threshold *max_level*. The detail of subdivision is presented in Section 3.4. Finally, the geometry of camera coverage is the union of grids whose encoding is not (0000); the area is also estimated.

3.2. Coverage Model for Ground Point

Two conditions need to be satisfied if a point is covered by a camera: the ray from the camera to the point should intersect with the image plane and there should not be an obstacle between the camera and the point. The former relates to the camera model and the latter to obstacles in the geographic environment. The camera model is illustrated in Figure 2. Camera C is located at (X_C, Y_C, H_C). Its coverage in theory is the pyramid $C\text{-}D_1D_2D_3D_4$, which is determined by intrinsic and external parameters of the camera. Intrinsic parameters include focal length f, principle center (u_0, v_0), etc. External parameters include pan angle P, tilt angle T, roll angle v, etc. P is the angle between the north direction and the principal optic axis in a clockwise direction, T is the angle from the horizontal direction to the principal optic axis in a clockwise direction, while v is often close to 0 and is ignored in this paper. The point G in the geographic environment is located at the coordinates (X_G, Y_G, H_G), the corresponding image point g, which is projected from point G by a camera, is located at the coordinates (x, y) in an image coordinate system. The camera model is shown as Equation (1), where λ is a non-zero scale factor:

$$\begin{bmatrix} f \\ x \\ y \end{bmatrix} = \begin{bmatrix} \cos T & 0 & -\sin T \\ 0 & 1 & 0 \\ \sin T & 0 & \cos T \end{bmatrix} \begin{bmatrix} \cos P & \sin P & 0 \\ -\sin P & \cos P & 0 \\ 0 & 0 & 1 \end{bmatrix} \begin{bmatrix} X_G - X_C \\ Y_G - Y_C \\ H_G - H_C \end{bmatrix}. \tag{1}$$

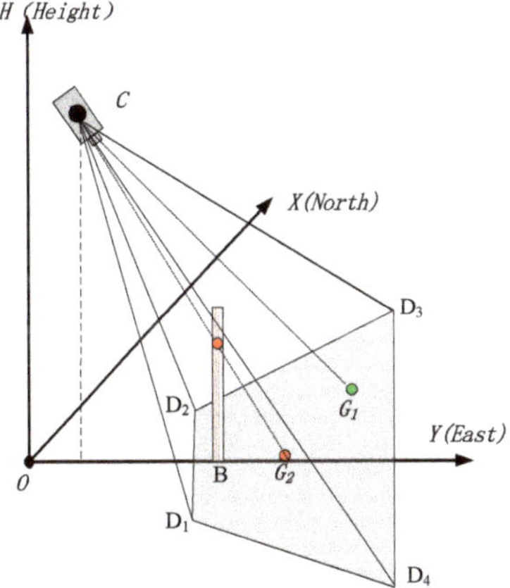

Figure 2. Camera model.

A point G is visible in an image if and only if the sight line CG determined by camera C and point G crosses the image plane and there is no obstacle across the sight line CG. As shown in Figure 2, the point G_1 is visible, but the point G_2 is blocked by obstacle B. The profile is shown in Figure 3. (X_B, Y_B, H_B) are the coordinates of B. The height H of the line of sight CG at the location of B is calculated by Equation (2):

$$H = \frac{l_2}{l_1 + l_2}H_G + \frac{l_1}{l_1 + l_2}H_C, \tag{2}$$

where $l_1 = \|(X_B - X_G, Y_B - Y_G)\|$, $l_2 = \|(X_B - X_C, Y_B - Y_C)\|$. If $H_B \geq H$, then the current point is visible. H_B can be obtained from the attribute tables of vector data.

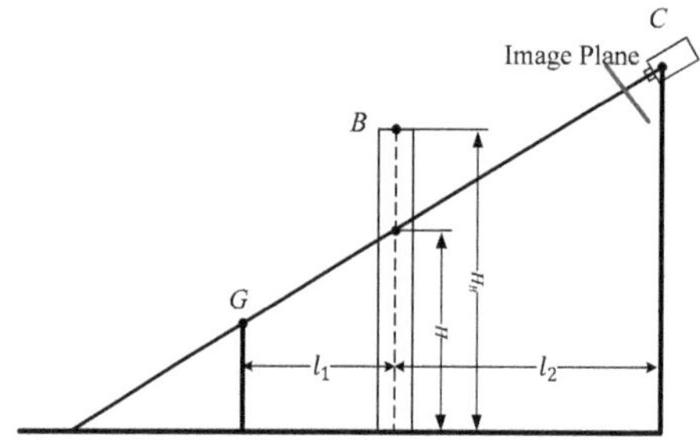

Figure 3. Profile of an object point, obstacle and camera.

3.3. Presentation for Grid

Each grid has four corners, so its status can be represented by four digits (0 or 1) according to their visibility. We arranged them in the left-up corner followed by right-up, left-down, and right-down. Consequently, there are 16 possibilities, which are illustrated as Table 1. If the status of the grid is (0000), the grid is not covered. If the status is (1111), the grid is covered. Other codes in the table represent partial coverages. As illustrated in Table 1, the codes (0110) and (1001) lead to ambiguity. Under the circumstances, an extra point should be sampled in the grid center to confirm the actual coverage.

Table 1. Codes of grids.

ID	Code	Coverage	ID	Code	Coverage	ID	Code	Coverage
0	0000					10	1010	
1	0001		6	0110		11	1011	
2	0010		7	0111		12	1100	
3	0011		8	1000		13	1101	
4	0100					14	1110	
5	0101		9	1001		15	1111	

3.4. Multistage Grid Subdivision

After dividing the MBR into unified grids, each grid needs to be reviewed to determine whether it should be subdivided further according to its status as presented in Section 3.3. For each grid in a level, there are two issues that need to be resolved: (a) convert and (b) conflict.

(a) convert

As shown in Figure 4, the quadrilateral with the blue border is the FOV of the camera in theory. The rectangle with a black bold border is its MBR. The positions of left-down and right-up points of the MBR are (*XMin*, *YMin*) and (*XMax*, *YMax*). The MBR is divided into grids with the initial grid size w_0. We record a corner point as (l, i, j), where l is the subdivision level of the points, i and j are the number of current corner points in the current subdivision level. A corner point (l, i, j) and its location in geographic coordinate (X, Y) can be converted from one to the other by following Equations (3) and (4):

$$
\begin{cases}
i = \left\lfloor \dfrac{(X - XMin) \times 2^l}{w_0} \right\rfloor \\
j = \left\lfloor \dfrac{(Y - YMin) \times 2^l}{w_0} \right\rfloor
\end{cases} , \tag{3}
$$

$$
\begin{cases}
X = \min(XMin + i \times \dfrac{w_0}{2^l}, XMax) \\
Y = \min(YMin + j \times \dfrac{w_0}{2^l}, YMax)
\end{cases} . \tag{4}
$$

The points (l, i, j) and $(l + 1, 2 \times i, 2 \times j)$ are located at the same place. Likewise, the points (l, i, j) and $(l + n, 2^n \times i, 2^n \times j)$ are the same point. When the grids are subdivided, only new points need to be estimated. Others can inherit their status from upper levels.

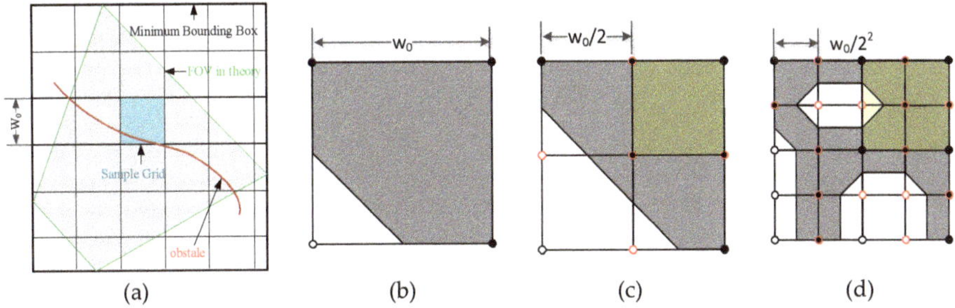

Figure 4. Different levels for grid-subdivision. (**a**) sample of FOV in theory; (**b**) sample grid ($l = 0$); (**c**) sample grid ($l = 1$); (**d**) sample grid ($l = 2$).

(b) conflict

The sample grid in blue shown in Figure 4a, whose status is (1101) in level 0 (see Figure 4b), is partly covered by the camera because of occlusion. Therefore, it needs further subdivision to the next level, which is shown in Figure 4c. The right-up grid masked in yellow in Figure 4c is coded as (1111); it does not need to be subdivided. However, its left neighbor grid needs to be subdivided, a new mid-point of the adjacent edge is added and its status is 0. This means that the grid masked in yellow must be subdivided because it is not completely covered by the camera. The current grid will be subdivided to the same level as its neighboring grid. Therefore, the grid is subdivided in Figure 4d. Likewise, if the status of the new mid-point is 1 and its neighboring grid is coded as (0000), then the neighboring grid will be subdivided.

Here is the algorithm (Algorithm 1):

Algorithms 1: Camera Coverage Estimation Based on Multistage Grid Subdivision

---

Input: Camera parameters, obstacle information, initial grid size w_0, max level *max_level*
Output: Geometry and area of coverage
Process:
    Subdivide the MBR of FOV in theory into grids with size w_0
    Set current subdivision level l to 0.
    While l < *max_level* and not all grids are coded as (0000) or (1111) do
        l++
    for each grid in level l
        Obtain and record the statuses of each grid (l, i, j).
        If its code is not (0000) or (1111), then
            detect the coverage statuses of five new points, which are composed of the center of the
        current grid, and the mid-points of four edges.
            record the statuses of each grid.
        For each new mid-point
            if it conflicts with the neighbor grid,
                then subdivide the neighbor grid to the current level..
    Convert the grid information from (l, i, j) to (X, Y)
    Generate the geometry of coverage, which is the union of all the grids in different levels.
    Obtain the area of coverage according to the geometry information.

---

4. Experiments and Results

The initial grid size and max level are two important factors that affect the accuracy and efficiency of the proposed method. To determine the impacts of the initial size and level of grid on the proposed method, a series of experiments were performed using simulated and real data.

In the experiments, we used the number of points needing to be judged for coverage by the camera to represent the efficiency of the method because the judgment process is the most time-consuming step. The more points that need to be judged, the more time-consuming the process. We employed the percentage of coverage area relative to real area to represent the accuracy of the simulated experiments.

4.1. Prototype System

Our method is designed for camera coverage estimation for the prototype system shown in Figure 5. The system is deployed in the sever with four main modules: (1) optimal camera network deployment, (2) camera control, (3) physical coverage visualization, and (4) spatial analysis for coverage. The system requires the accurate geometry of the individual cameras and camera network for coverage visualization and spatial analysis, and the acceptable speed to obtain optimal deployment scheme. Only the certain camera parameters need to be transferred between system and the corresponding camera other than camera coverage. The communication complexity is out of range of our method. Consequently, coverage estimation method considering the trade-off of efficacy and accuracy is desirable.

The experimental environment of this study is Ubuntu 64-bit operating system, Intel i5 processor, 2.0 G memory (San Jose, CA, USA, Apple). The study uses Python as the developing language, an open source QGIS to carry out geometric target description and topological relations operation.

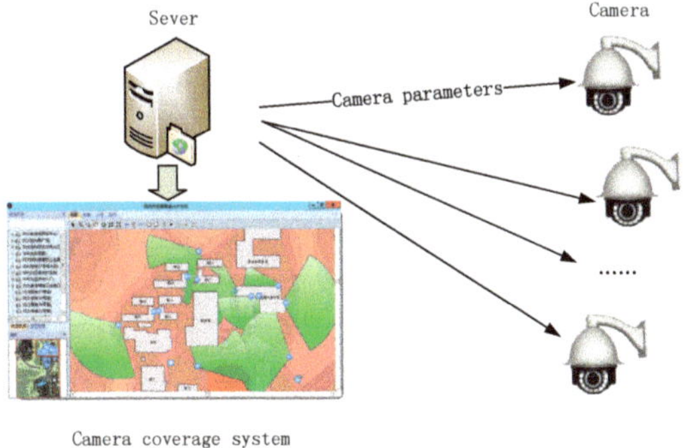

Figure 5. System prototype.

4.2. Simulated Data

In this section, we employed three geometrical objects to simulate different geographic environments with different complexity. Three geometric shapes are covered by a camera, and the other areas covered by the camera are ignored because the process for them is the same for our method as for others. We employed a circle with a radius of 100 units, a diamond with side length of 100 units, and a five-pointed star with external and internal radiuses of 100 and 50 units to simulate different coverage situations. The circle is the simplest one, while the five-pointed star is the most complex.

As shown in Figures 6 and 7, the red area is the real coverage and the blue area is obtained by the proposed method with different initial grid sizes and max levels. Points filled with white mean the corner points are not covered by a camera while the ones filled with black mean that they are covered. In Figure 6, the max level is set to 1, and the initial grid sizes are specified as 100, 75, 50, 25 and 5. Similarly, in Figure 7, the initial grid sizes are set to 100, and the max level for subdivision is specified as 1, 2, 3 and 4.

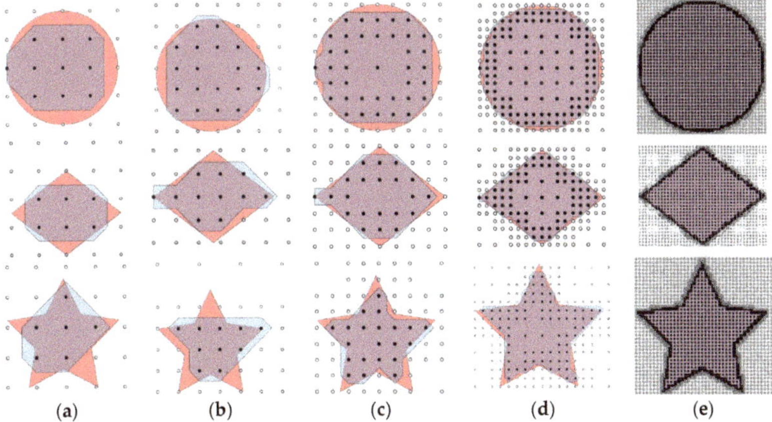

Figure 6. Different initial grid sizes for the proposed method with max level = 1. (**a**) initial grid size = 100; (**b**) initial grid size = 75; (**c**) initial grid size = 50; (**d**) initial grid size = 25; (**e**) initial grid size = 5.

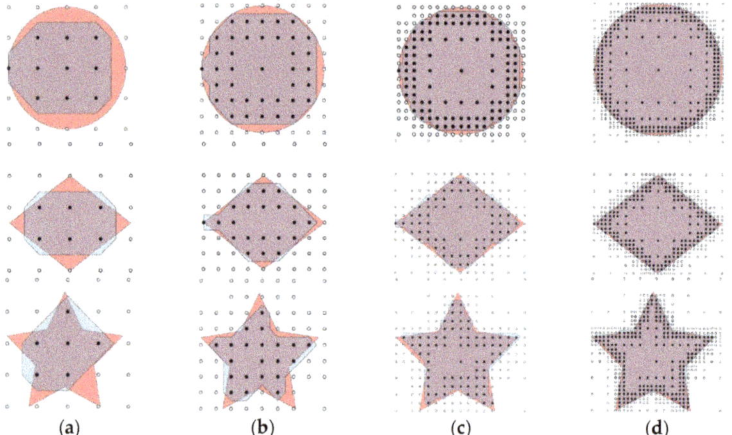

Figure 7. Different max levels for the proposed method with initial grid size = 100. (**a**) level = 1; (**b**) level = 2; (**c**) level = 3; (**d**) level = 4.

The results of experiments with different initial grid sizes and max levels are shown in Figures 8–10. In these figures, the ***point number*** stands for efficiency, which is represented by the number of points needing to be judged for whether they are covered by the camera. The ***coverage rate*** stands for accuracy, which is represented by the percentage of the coverage area relative to the real area.

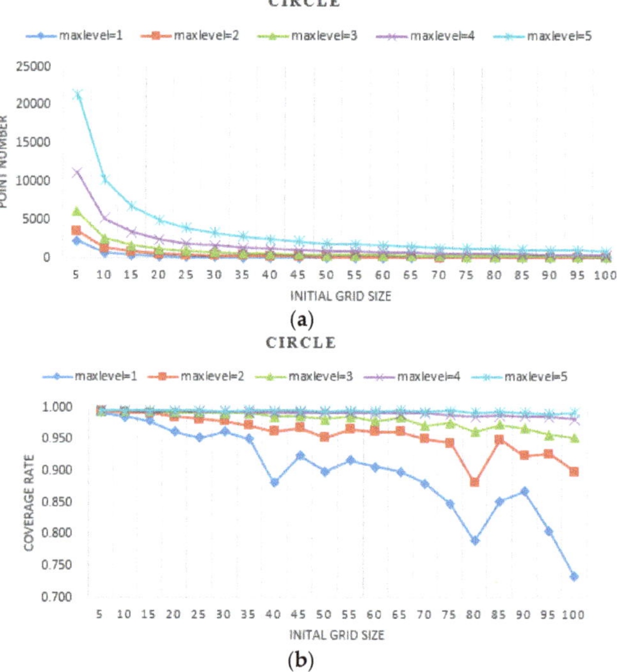

Figure 8. Results for the circle with different initial grid sizes and max levels. (**a**) point number; (**b**) coverage rate.

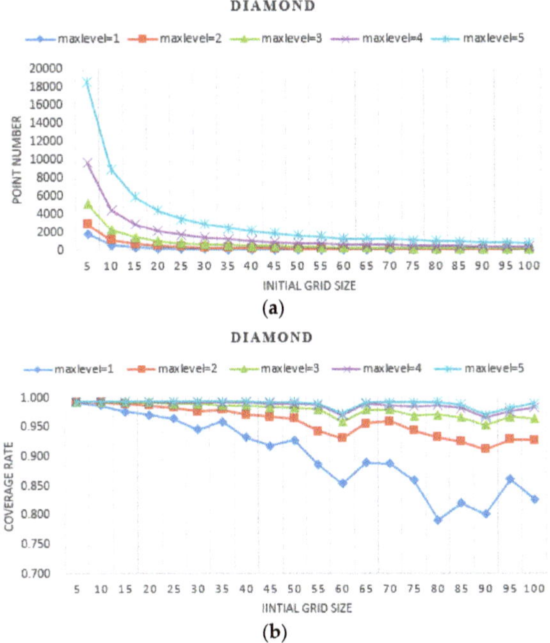

Figure 9. Results for the diamond with different initial grid sizes and max levels. (**a**) point number; (**b**) coverage rate.

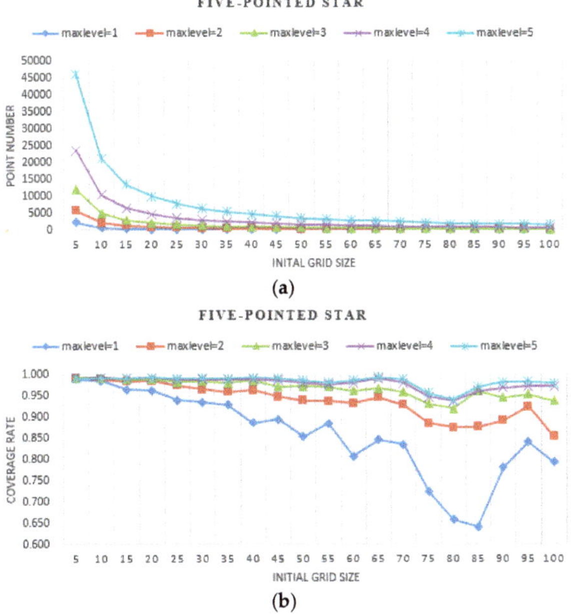

Figure 10. Results for the five-pointed star with different initial grid sizes and max levels. (**a**) point number; (**b**) coverage rate.

From the details illustrated in Figures 6–10, the following considerations can be remarked.

(a)   When the initial grid size is fixed, as the max level increases, the geometries of the simulated shapes are closer to the real shapes, and the point number of the proposed method increases dramatically. There are twice the point numbers of the former max level, and as the max level increases accuracy increases.

(b)   When the max level is fixed, as the initial grid size increases, the geometries of the simulated shapes are closer to the real shapes, and the point number of the proposed method decreases. At a small initial grid size, the number of points declined sharply and leveled off gradually with the increase of the initial grid size. As the initial grid size increased, the coverage rate decreased overall. The larger the max level is, the slower the coverage rate decreases.

(c)   When the initial grid size is small, in the experiments, it is set to 5, and the accuracy of the proposed method for all three shapes is high, approaching 99%. The point number increases dramatically as the max level increases.

(d)   When the initial grid size is large, in the experiments, it is larger than half the shape width, and the accuracy of the proposed method for all three shapes is slightly unstable, but it decreases overall. The point numbers become close to each other.

(e)   The point numbers for the five-pointed star are more than the other two shapes, and the coverage rate is a little less with the same initial grid size and max level. Because the five-pointed star simulated the complex geographic phenomenon, most of the grids needed to be subdivided.

(f)   The coverage rate vibrates, which is shown in Figures 8–10. The points filled in red, blue and green in Figure 11 are the grid points, which the FOV is divided into with the certain initial grid size of 25, 15 and 10. In addition, the corresponding sub-grid points are filled in with similar colors. As shown in Figure 11, the sub-grid points need to be judged with initial grid size of 25, 15 and 10 not overlapping. Consequently, the status of each grid point is not the same, and then the coverage rate vibrates as shown in Figures 8–10.

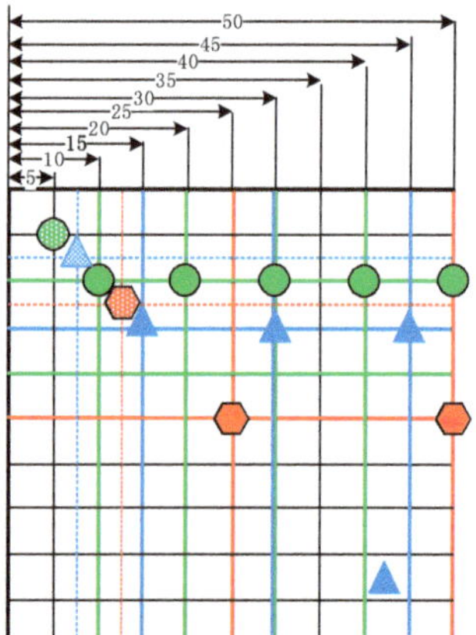

Figure 11. Initial grid size and subdivision.

4.3. Real Geographic Environment Data

As illustrated in Figure 12, there are 15 cameras deployed, including eight PTZ (Pan/Tilt/Zoom) ones and seven still ones. PTZ cameras can rotate and tilt at a certain angle and provide optical zoom; therefore, their coverage is a sector composed of the coverages from all possible camera positions. In the experiment, the steps for pan and tile are one degree. If the pan and tile range are (230,310) and (25, 65) respectively, then the coverage is estimated 3200 times. Consequently, the point numbers of

PTZ cameras is the sum of point numbers from cameras with certain pan and tile. The still camera's coverage is a quadrangle. All the camera parameters are listed in Table 2. Their locations and coverages are illustrated in Figure 12. The digits in red represent the ID of the camera, and the areas in transparent blue are the coverages estimated by our proposed method. In the experiment, the buildings are the major obstacles because the height of the cameras is much lower than building height. There are 85 features in the building layer.

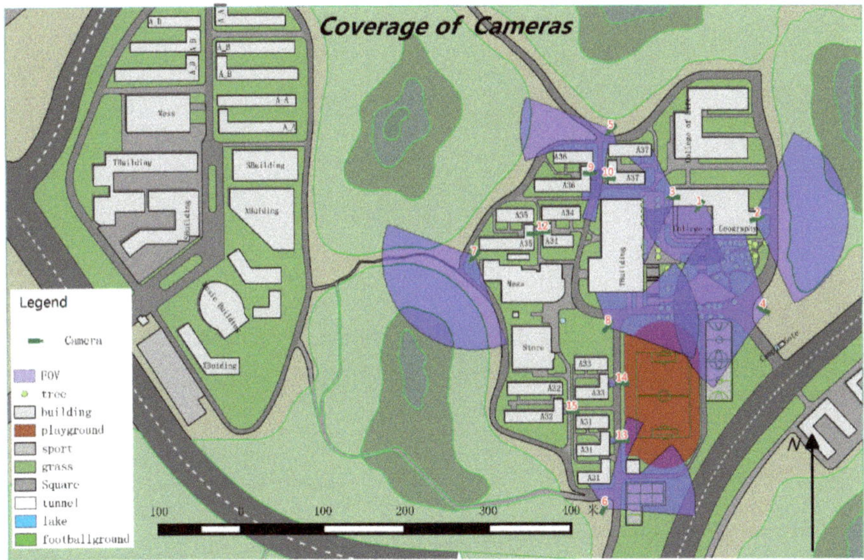

Figure 12. Cameras' coverage in geographic environment.

Table 2. Camera parameters.

Type	Id	Height (Meter)	Format (Millimeter × Millimeter)	Focal (Millimeter)	Pan (Degree)		Tilt (Degree)	
					Mini	Max	Mini	Max
PTZ Camera	1	14	4.8 × 3.6	3.6	230	310	25	65
	2	14	4.8 × 3.6	3.6	163	243	46	86
	3	16.8	4.8 × 3.6	3.6	50	130	25	65
	4	14	4.8 × 3.6	3.6	240	320	20	60
	5	6	4.8 × 3.6	3.6	185	265	20	60
	6	6	4.8 × 3.6	3.6	350	70	20	60
	7	6	4.8 × 3.6	3.6	185	265	20	60
	8		4.8 × 3.6	3.6	5	85	20	60
Still Camera	9	3	3.2 × 2.4	3.6	90		80	
	10	3	3.2 × 2.4	3.6	280		80	
	11	3	3.2 × 2.4	3.6	90		80	
	12	3	3.2 × 2.4	3.6	270		80	
	13	3	3.2 × 2.4	3.6	270		80	
	14	3	3.2 × 2.4	3.6	270		80	
	15	3	3.2 × 2.4	3.6	270		80	

In this experiment, we first set the initial grid size to 4, 2, 1 and 0.5 m. Then, we estimated camera coverages without further subdivision. Second, we set the initial grid size to 4 m and set the max level to 0, 1, 2 and 3. Third, we set the size of the grid max level to 0.5 m. In other words, we set the initial size and max level as 4 m and three levels, 2 m and two levels, 1 m and one level, and 0.5 m without subdivision. Because of ignorance of the ground truth, we compared our estimated results with ones taking 0.5 m as the initial grid size and 0 level as the max level. Because of differences in the order of magnitude, the results of PTZ cameras and still cameras are illustrated in Figures 13 and 14, respectively.

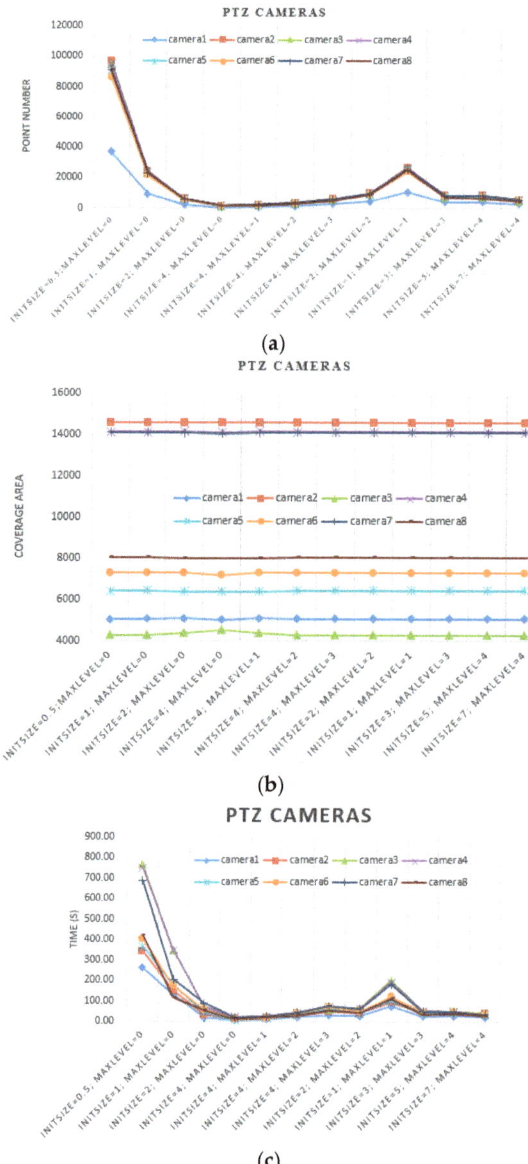

Figure 13. Results for PTZ cameras (**a**) point number; (**b**) coverage area; (**c**) time-consuming.

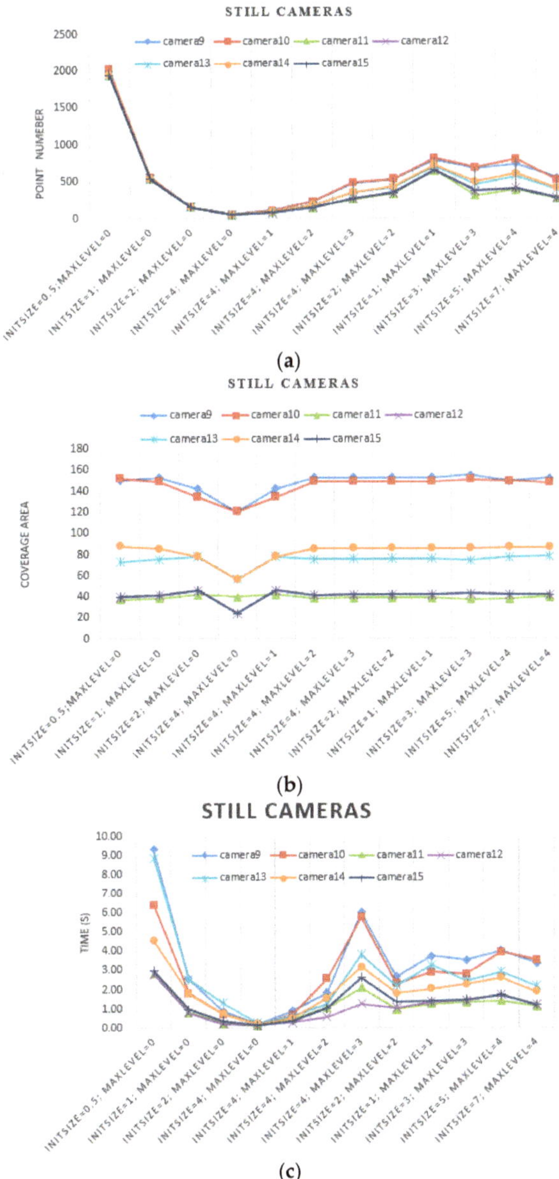

Figure 14. Results for still cameras (**a**) point number; (**b**) coverage area; (**c**) time-consuming.

From the results illustrated in Figures 13 and 14, the same conclusions can be made as with the experiment with simulated data, along with the following considerations:

(a) When the size of the grid in max level is the same, which is 0.5 m for example, the initial size and max level are set as 4 m and three levels, 2 m and two levels, 1 m and one level, the point numbers increase with the initial grid size, and they are significantly lower than results with

0.5 m as the initial grid size and 0 as the max level. However, the coverage areas are close to the ground truth.

(b)   When the size of the grid in max level is similar, for example, the initial size and max level are set as 3 m and three levels, 5 m and four levels, 7 m and four levels, the point numbers and coverage area are close to each other.

(c)   On one hand, the point number depends on the camera's physical coverage, which is influenced by camera parameters and geographic environment. As the physical coverage increases, the point number increases. On the other hand, the point number is influenced by the initial grid size together with the max level proposed by our method.

(d)   As shown in Figures 13c and 14c, with the same initial grid size, the processing time of different cameras increases as the max level increases. With the same max level, the processing time increases with the initial grid size. When the size of the grid at the max level is the same, the processing times of different cameras are close to each other. Even though the point numbers of different cameras are close to each other, the processing times vary. Moreover, the processing times of different cameras vary because of their locations, poses and obstacles.

(e)   The processing times of PTZ cameras is very time-consuming because the total coverage is combined with lots of coverages estimated with certain pan and tile.

5. Analysis and Discussion

The accuracy and efficiency of our proposed method are greatly influenced by camera parameters, obstacles, initial grid size and max level. The camera parameters can be employed to estimate the FOV in the theory, and obstacles must be considered when physical coverage is needed. However, it is hard to make a quantitative analysis of the influences before camera deployment. In general, cameras for city public security are usually deployed in entrances, exits and road intersections for monitoring moving targets. The geographic environment with obstacles such as buildings and trees is simpler than the simulated five-star. Consequently, in the paper, we emphasized the later factors: the initial grid size and the max level.

We use N_l and M_l to represent the row and column number of grid points from subdivision of the MBR with unified grid size $w_0/2^l$, where w_0 is the initial grid size, and l is the current subdivision level:

$$\begin{cases} N_l = \left\lceil \frac{(XMax-XMin)\times 2^l}{w_0} \right\rceil + 1 \\ M_l = \left\lceil \frac{(YMax-YMin)\times 2^l}{w_0} \right\rceil + 1 \end{cases}. \tag{5}$$

Therefore, the number of grid points from subdivision of the MBR with unified grid size w_0, which is written as $GridPointNum_0$, is computed by Equation (5) with $l = 0$. The number of grid points from subdivision of the MBR with unified grid size $w_0/2^l$ is written as $GridPointNum_l$ in Equation (6). In theory, the number grid points should be estimated in the proposed method with initial grid size w_0 and max level l, which is written as $GridPointNum_{w_0\_l}$, not less than $GridPointNum_0$ and not bigger than $GridPointNum_l$. That is, $GridPointNum_{w_0\_l} \in [GridPointNum_0, GridPointNum_l]$:

$$\begin{cases} GridPointNum_0 = N_0 \times M_0 \\ GridPointNum_l = N_l \times M_l \end{cases}. \tag{6}$$

Consequently, the time complexity of our algorithm is $O(GridPointNum_{w_0\_l} \times featureNum)$. To avoid judging the status of gird point repeatedly, our method needs to record the judged grid points. Consequently, the space complexity is also $O(GridPointNum_{w_0\_l})$. In reality, when the camera is deployed in an environment with complex occlusions, the efficacy of the proposed method is close to $GridPointNum_l$. When the camera is deployed in a relatively flat area with few obstacles, the proposed method is more efficient.

On average, our method trades off efficacy and accuracy. Experiments with simulated and real data reveal the same conclusions. Overall, the oversize initial grid results in less accuracy, and the oversize max level is less efficient without obvious accuracy improvement. An undersize initial grid results in more computing time, and the undersize max level could cause less accuracy. Consequently, it is important to choose a proper combination of the initial grid size and max level. In application, there are three suggestions resulting from our experiments:

(1) If high efficacy is given priority over high accuracy, a larger initial grid size and smaller max level should be chosen.
(2) If high accuracy is given priority over high efficacy, a smaller initial grid size and larger max level is appropriate.
(3) When the focus is a balance between accuracy and efficacy, the parameters can be determined by the following steps: (a) roughly estimate the FOVs in theory and their MBRs; (b) estimate the smallest grid size and max level for the desired accuracy; and (c) estimate the initial grid size and max level for acceptable efficacy and accuracy using Equations (5) and (6).

In this paper, there are some limitations. This is unavoidable when sampling. In theory, if the grid size is small enough, a best grid approximation will be obtained, but it is impractical to divide the area infinitely. It is usually divided into grids according to practical requirements.

(1) If the initial grid size is not small enough, our method may ignore the conditions, which are illustrated in Figure 15. When the grid coded as (1111) has a few holes, its geometry and area are overestimated. When the grid coded as (0000) has a few islands, its geometry and area are underestimated. To avoid or reduce the impacts of sampling without loss of computing efficiency, it is suitable to choose a relatively smaller initial grid size and then determine the max level according to the desired deepest grid size. The conditions shown in Figure 15 are infrequent.

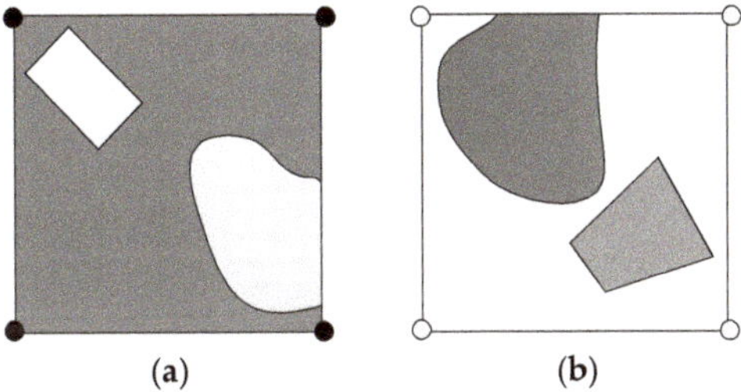

(a)   (b)

Figure 15. Exceptions for grid subdivision. (a) holes; (b) islands.

(2) The efficacy and accuracy of our method is affected by boundaries. As shown in Figures 5 and 6, the boundary of physical coverage is not perpendicular to the vertical or horizontal direction. Therefore, the estimated coverage is serrated, and the grids crossing the boundary need to be divided by the max level to approach the physical coverage, which may cause more computing time.

(3) The monitored area in our method is flat ground, some errors may result when the area is rolling, and a few points may be occluded by the terrain. Our method can be improved for 3D terrain because its core for a visibility test is LOS when high precision DEM/DSM is accessible.

6. Conclusions

In this paper, a method is proposed to estimate camera coverage that balances accuracy and efficacy. In this method, the camera FOV in theory is divided by grids of different sizes with on-demand accuracy rather than by grids with one fixed size. Accuracy is approximately equivalent to the method employing the same deepest grid size, but efficacy is equivalent to the method employing the same initial grid size. It is suitable for a camera network, which contains hundreds of cameras and needs to obtain coverage frequently because of reconfiguration, coverage enhancement, optimal placement, etc. In this paper, we employed the LOS to estimate the visibility of the grid corner points. Even though the experiments cater to 2D areas with obstacles in vector format, it is easy to expand to 3D camera coverage when the high-precision grid DEM is available. In addition, different LODs of 3D buildings will be considered in our future works.

Acknowledgments: The work described in this paper was supported by the National Natural Science Foundation of China (NSFC) (Grant No. 41401442), the National High Technology Research and Development Program of China (Grant No. 2015AA123901), the National Key Research and Development Program (Grant No. 2016YFE0131600), the Sustainable Construction of Advantageous Subjects in Jiangsu Province (Grant No. 164320H116) and the Priority Academic Program Development of Jiangsu Higher Education Institutions.

Author Contributions: Wang Meizhen designed the methodology of research, coordinated the implementation of the proposed approach and wrote the paper. Liu Xuejun outlined the framework of this paper. Zhang Yanan contributed to implementing the approach. Wang Ziran contributed to implementing the approach and preparing the illustrative figures.

Conflicts of Interest: The authors declare no conflicts of interest.

References

1. Mavrinac, A.; Chen, X. Modeling coverage in camera networks: A survey. *Int. J. Comput. Vis.* **2013**, *101*, 205–226. [CrossRef]
2. Han, Z.; Cui, C.; Kong, Y.; Qin, F.; Fu, P. Video data model and retrieval service framework using geographic information. *Trans. GIS* **2016**, *20*, 701–717. [CrossRef]
3. Lewis, P.; Fotheringham, S.; Winstanley, A.C. Spatial video and GIS. *Int. J. Geogr. Inf. Sci.* **2011**, *25*, 697–716. [CrossRef]
4. Milosavljevic, A.; Dimitrijevic, A.; Rancic, D. GIS-augmented video surveillance. *Int. J. Geogr. Inf. Sci.* **2010**, *24*, 1415–1433. [CrossRef]
5. Milosavljevic, A.; Rancic, D.; Dimitrijevic, A.; Predic, B.; Mihajlovic, V. Integration of GIS and video surveillance. *Int. J. Geogr. Inf. Sci.* **2016**, *30*, 2089–2107. [CrossRef]
6. Zhang, X.; Liu, X.; Song, H. Video surveillance GIS: A novel application. In Proceedings of the 2013 21st International Conference on Geoinformatics, Kaifeng, China, 20–22 June 2013; pp. 1–4.
7. Fu, Y.; Zhou, J.; Deng, L. Surveillance of a 2D plane area with 3D deployed cameras. *Sensors* **2014**, *14*, 1988–2011. [CrossRef] [PubMed]
8. Hörster, E.; Lienhart, R. On the optimal placement of multiple visual sensors. In Proceedings of the 4th ACM International Workshop on Video Surveillance and Sensor Networks, Santa Barbara, CA, USA, 27 October 2006; pp. 111–120.
9. Erdem, U.M.; Sclaroff, S. Automated camera layout to satisfy task-specific and floor plan-specific coverage requirements. *Comput. Vis. Image Underst.* **2006**, *103*, 156–169. [CrossRef]
10. Tao, D.; Ma, H.; Liu, L. Coverage-enhancing algorithm for directional sensor networks. In Proceedings of the International Conference on Mobile Ad-Hoc and Sensor Networks (MSN 2006), Hong Kong, China, 13–15 December 2006; Volume 4325, pp. 256–267.
11. Ma, H.; Liu, Y. On coverage problems of directional sensor networks. In Proceedings of the International Conference on Mobile Ad-hoc and Sensor Networks (MSN 2005), Wuhan, China, 13–15 December 2005; Volume 3794, pp. 721–731.
12. Ma, H.; Zhang, X.; Ming, A. A coverage-enhancing method for 3D directional sensor networks. In Proceedings of the International Conference on Computer Communications, Rio de Janeiro, Brazil, 19–25 April 2009.
13. Xiao, F.; Wang, R.-C.; Sun, L.-J.; Wu, S. Research on the three-dimensional perception model and coverage-enhancing algorithm for wireless multimedia sensor networks. *J. China Univ. Posts Telecommun.* **2010**, *17*, 67–72. [CrossRef]
14. Xu, Y.-C.; Lei, B.; Hendriks, E.A. Camera network coverage improving by particle swarm optimization. *J. Image Video Process.* **2011**, *2011*, 458283. [CrossRef]

15. Jiang, P.; Jin, W. A PSO-based algorithm for video networks planning optimization. In Proceedings of the 2013 6th International Congress on Image and Signal Processing (CISP), Hangzhou, China, 16–18 December 2013; pp. 122–126.
16. Zhao, J.; Cheung, S.S. Optimal visual sensor planning. In Proceedings of the International Symposium on Circuits and Systems, Taipei, Taiwan, 24–27 May 2009.
17. Qian, C.; Qi, H. Coverage estimation in the presence of occlusions for visual sensor networks. In Proceedings of the Distributed Computing in Sensor Systems (DCOSS 2008), Santorini Island, Greece, 11–14 June 2008.
18. Karakaya, M.; Qi, H. Coverage estimation in heterogeneous visual sensor networks. In Proceedings of the Distributed Computing in Sensor Systems, Hangzhou, China, 16–18 May 2012.
19. Angella, F.; Reithler, L.; Gallesio, F. Optimal deployment of cameras for video surveillance systems. In Proceedings of the Advanced Video and Signal Based Surveillance, London, UK, 5–7 September 2007.
20. Debaque, B.; Jedidi, R.; Prevost, D. Optimal video camera network deployment to support security monitoring. In Proceedings of the 2009 12th International Conference on Information Fusion, Seattle, WA, USA, 6–9 July 2009; pp. 1730–1736.
21. SanMiguel, J.C.; Micheloni, C.; Shoop, K.; Foresti, G.L.; Cavallaro, A. Self-reconfigurable smart camera networks. *Computer* **2014**, *47*, 67–73.
22. Zhou, P.; Long, C. Optimal coverage of camera networks using PSO algorithm. In Proceedings of the 2011 4th International Congress on Image and Signal Processing, Shanghai, China, 15–17 October 2011; pp. 2084–2088.
23. Chen, T.-S.; Tsai, H.-W.; Chen, C.-P.; Peng, J.-J. Object coverage with camera rotation in visual sensor networks. In Proceedings of the 6th International Wireless Communications and Mobile Computing Conference, Caen, France, 28 June–2 July 2010; pp. 79–83.
24. Munishwar, V.P.; Abu-Ghazaleh, N.B. Scalable target coverage in smart camera networks. In Proceedings of the Fourth ACM/IEEE International Conference on Distributed Smart Cameras, Atlanta, GA, USA, 31 August–4 September 2010; pp. 206–213.
25. Dieber, B.; Micheloni, C.; Rinner, B. Resource-aware coverage and task assignment in visual sensor networks. *IEEE Trans. Circuits Syst. Video Technol.* **2011**, *21*, 1424–1437. [CrossRef]
26. Song, X.; Tang, G.; Liu, X.; Dou, W.; Li, F. Parallel viewshed analysis on a PC cluster system using triple-based irregular partition scheme. *Earth Sci. Inf.* **2016**, *9*, 1–13. [CrossRef]
27. Wang, J.; Robinson, G.J.; White, K. Generating viewsheds without using sightlines. *Photogramm. Eng. Remote Sens.* **2000**, *66*, 87–90.
28. Kaucic, B.; Zalik, B. Comparison of viewshed algorithms on regular spaced points. In Proceedings of the Spring Conference on Computer Graphics, San Diego, CA, USA, 21–23 June 2002.
29. De Floriani, L.; Magillo, P. Algorithms for visibility computation on terrains: A survey. *Environ. Plan. B Plan. Des.* **2003**, *30*, 709–728. [CrossRef]
30. Fang, C.; Yang, C.; Chen, Z.; Yao, X.; Guo, H. Parallel algorithm for viewshed analysis on a modern GPU. *Int. J. Digita. Earth* **2011**, *4*, 471–486.
31. Cauchisaunders, A.J.; Lewis, I. GPU enabled XDraw viewshed analysis. *J. Parallel Distrib. Comput.* **2015**, *84*, 87–93. [CrossRef]
32. Wang, F.; Wang, G.; Pan, D.; Yuan, L.; Yang, L.; Wang, H. A parallel algorithm for viewshed analysis in three-dimensional digital earth. *Comput. Geosci.* **2015**, *75*, 57–65.
33. Yaagoubi, R.; Yarmani, E.M.; Kamel, A.; Khemiri, W. Hybvor: A voronoi-based 3D GIS approach for camera surveillance network placement. *ISPRS Int. J. Geo-Inf.* **2015**, *4*, 754–782. [CrossRef]
34. Liu, L.; Ma, H. On coverage of wireless sensor networks for rolling terrains. *IEEE Trans. Parallel Distrib. Syst.* **2012**, *23*, 118–125. [CrossRef]
35. Argany, M.; Mostafavi, M.A.; Akbarzadeh, V.; Gagne, C.; Yaagoubi, R. Impact of the quality of spatial 3D city models on sensor networks placement optimization. *Geoinformatica* **2012**, *66*, 291–305. [CrossRef]
36. ArcGIS Help 10.1, Viewshed (3D Analyst). Available online: http://resources.arcgis.com/en/help/main/10.1/index.html#//00q900000033000000 (accessed on 5 October 2016).

MDPI
St. Alban-Anlage 66
4052 Basel
Switzerland
Tel. +41 61 683 77 34
Fax +41 61 302 89 18
www.mdpi.com

ISPRS International Journal of Geo-Information Editorial Office
E-mail: ijgi@mdpi.com
www.mdpi.com/journal/ijgi